T0202760

Communications in Computer and Information Science 1866

Rationale

The CCIS series is devoted to the publication of proceedings of computer science conferences. Its aim is to efficiently disseminate original research results in informatics in printed and electronic form. While the focus is on publication of peer-reviewed full papers presenting mature work, inclusion of reviewed short papers reporting on work in progress is welcome, too. Besides globally relevant meetings with internationally representative program committees guaranteeing a strict peer-reviewing and paper selection process, conferences run by societies or of high regional or national relevance are also considered for publication.

Topics

The topical scope of CCIS spans the entire spectrum of informatics ranging from foundational topics in the theory of computing to information and communications science and technology and a broad variety of interdisciplinary application fields.

Information for Volume Editors and Authors

Publication in CCIS is free of charge. No royalties are paid, however, we offer registered conference participants temporary free access to the online version of the conference proceedings on SpringerLink (http://link.springer.com) by means of an http referrer from the conference website and/or a number of complimentary printed copies, as specified in the official acceptance email of the event.

CCIS proceedings can be published in time for distribution at conferences or as post-proceedings, and delivered in the form of printed books and/or electronically as USBs and/or e-content licenses for accessing proceedings at SpringerLink. Furthermore, CCIS proceedings are included in the CCIS electronic book series hosted in the SpringerLink digital library at http://link.springer.com/bookseries/7899. Conferences publishing in CCIS are allowed to use Online Conference Service (OCS) for managing the whole proceedings lifecycle (from submission and reviewing to preparing for publication) free of charge.

Publication process

The language of publication is exclusively English. Authors publishing in CCIS have to sign the Springer CCIS copyright transfer form, however, they are free to use their material published in CCIS for substantially changed, more elaborate subsequent publications elsewhere. For the preparation of the camera-ready papers/files, authors have to strictly adhere to the Springer CCIS Authors' Instructions and are strongly encouraged to use the CCIS LaTeX style files or templates.

Abstracting/Indexing

CCIS is abstracted/indexed in DBLP, Google Scholar, EI-Compendex, Mathematical Reviews, SCImago, Scopus. CCIS volumes are also submitted for the inclusion in ISI Proceedings.

How to start

To start the evaluation of your proposal for inclusion in the CCIS series, please send an e-mail to ccis@springer.com.

Mukesh Kumar Saini · Neeraj Goel ·
Hanumant Singh Shekhawat ·
Jaime Lloret Mauri · Dhananjay Singh
Editors

Agriculture-Centric Computation

First International Conference, ICA 2023
Chandigarh, India, May 11–13, 2023
Revised Selected Papers

 Springer

Editors
Mukesh Kumar Saini (iD)
Indian Institute of Technology Ropar
Ropar, India

Neeraj Goel (iD)
Indian Institute of Technology Ropar
Ropar, India

Hanumant Singh Shekhawat (iD)
Indian Institute of Technology Guwahati
Guwahati, India

Jaime Lloret Mauri (iD)
Polytechnic University of Valencia
Valencia, Spain

Dhananjay Singh (iD)
Saint Louis University
St. Louis, MO, USA

ISSN 1865-0929 ISSN 1865-0937 (electronic)
Communications in Computer and Information Science
ISBN 978-3-031-43604-8 ISBN 978-3-031-43605-5 (eBook)
https://doi.org/10.1007/978-3-031-43605-5

This Springer imprint is published by the registered company Springer Nature Switzerland AG
The registered company address is: Gewerbestrasse 11, 6330 Cham, Switzerland

Paper in this product is recyclable.

Preface

Agriculture has been the foundation of human civilization for thousands of years, and with the world population expected to reach 9.7 billion by 2050, it is more important than ever to find sustainable solutions to feed the Earth. As our population continues to grow, the demand for food and the need for innovative solutions in agriculture are increasing. At the same time, advances in science and computing technology have opened new possibilities for precision agriculture, data-driven decision-making, and automation in the last few years.

ICA 2023 is one of the rare conferences focusing on "Agriculture-Centric Computation", which explores the exciting and rapidly evolving intersection of agriculture and computing. This integration has the potential to revolutionize the way we produce, distribute, and consume food. It is a privilege to present the proceedings of the International Conference on Agriculture-Centric Computation (ICA 2023), held during May 11–13, 2023 at the Indian Institute of Technology Ropar (Punjab), India. The conference is funded by Agriculture and Water Technology Development Hub (iHub– AWaDH), which is established by the Department of Science & Technology (DST), Government of India, at the Indian Institute of Technology Ropar.

ICA is a platform where we explore the research challenges emerging in the complex interaction between computing and agriculture. We examine how computing can be applied to agriculture to address some of the biggest challenges facing the industry today, such as climate change, food security, and environmental sustainability. We delve into topics such as big data analytics, artificial intelligence, machine learning, the Internet of Things (IoT), remote sensing, robotics, and drones, and how they can be used to optimize crop yields, reduce resource consumption, and improve farm profitability. All the stakeholders, such as leading researchers, practitioners, and industry representatives, participate and share their views.

Out of 52 submitted papers, 18 were accepted for oral presentation and publication by the program committee based on the recommendations of at least 2 expert reviewers in a double-blind review process. ICA 2023 included three keynote speakers and four invited talks, panel discussions, interaction with progressive farmers, twenty poster presentations, and nineteen powerful expert session chairs who have worked in both industry and academia.

ICA attracts gatherings of academic researchers, undergraduate students, postgraduate students, research scholars, top research think tanks, and industry technology developers. Therefore, we do believe that the biggest benefit to the participant is the actualization of their goals in the field of agriculture-centric computation. That will ultimately lead to greater success in agriculture, which is beneficial to society. Moreover, our warm gratitude should be extended to all the authors who submitted their work to ICA 2023. During the submission, review, and editing stages, the EasyChair conference system proved very helpful. We are grateful to the technical program committee (TPC) and the local organizing committee for their immeasurable efforts to ensure the success of this conference. Finally, we would like to thank our speakers, authors, and participants for their contributions to making ICA 2023 a stimulating and productive conference.

May 2023

Mukesh Kumar Saini
Neeraj Goel
Hanumant Singh Shekhawat
Jaime Lloret Mauri
Dhananjay Singh

Organization

Patron

Rajeev Ahuja   IIT Ropar, India

General Chairs

Dhananjay Singh   Saint Louis University, USA
Abdulmotaleb El Saddik   MBZUAI, UAE & University of Ottawa, Canada

Technical Program Chairs

Mukesh Kumar Saini   IIT Ropar, India
Neeraj Goel   IIT Ropar, India
Hanumant Singh Shekhawath   IIT Guwahati, India
Jaime Lloret Mauri   Valencia Polytechnic University, Spain
Dhananjay Singh   Saint Louis University, USA

Organizing Committee

Dhananjay Singh   Saint Louis University, USA
Abdulmotaleb El Saddik   MBZUAI, UAE & University of Ottawa, Canada
Mukesh Kumar Saini   IIT Ropar, India
Neeraj Goel   IIT Ropar, India
Hanumant Singh Shekhawat   IIT Guwahati, India
Jaime Lloret Mauri   Valencia Polytechnic University, Spain
Pradeep Atrey   State University of New York at Albany, USA
Pushpendra Pal Singh   IIT Ropar, India
Ekta Kapoor   DST, Government of India, India
Monisa Qadiri   IUST, India
Swati Shukla   VIT-AP University, India
Pushpendra Pal Singh   IIT Ropar, India
Rohitashw Kumar   SKUAST-Kashmir, India
Deepika Gupta   Galgotias University, India

Anterpreet Bedi	Thapar Institute of Engineering & Technology, India
Dipika Deb	Intel Corporation, India
Simrandeep Singh	IIT Ropar, India
Mangal Kothari	IIT Kanpur, India
Gayatri Ananthnarayanan	IIT Dharwad, India
T. V. Kalyan	IIT Ropar, India
Pratibha Kumari	University of Regensburg, Germany
Suman Kumar	IIT Ropar, India
Yamuna Prasad	IIT Jammu, India

Advisory Board

Pradeep Atrey	State University of New York at Albany, USA
Pushpendra Pal Singh	IIT Ropar, India
Ekta Kapoor	DST, Government of India, India

Workshop and Tutorial Chairs

Swati Shukla	VIT-AP University, India
Deepika Gupta	Galgotias University, India

Demo Chairs

Anterpreet Kaur Bedi	Thapar Institute of Engineering & Technology, India
Dipika Deb	Intel Corporation, India

Sponsorship and Industry Engagement Chairs

Suman Kumar	IIT Ropar, India
Pushpendra Pal Singh	IIT Ropar, India

Award Chairs

Mangal Kothari	IIT Kanpur, India
Gayatri Ananthnarayanan	IIT Dharwad, India

Registration Chairs

T. V. Kalyan   IIT Ropar, India
Yamuna Prasad   IIT Jammu, India

Proceedings Chair

Rohitashw Kumar   SKUAST-Kashmir, India

Web Chair

Hanumant Singh Shekhawat   IIT Guwahati, India

Media Chair

Monisa Qadiri   IUST, India

Keynote Speakers

Abdulmotaleb El Saddik   MBZUAI, UAE & University of Ottawa, Canada
Amy Reibman   Purdue University, USA
Athula Ginige   Western Sydney University, Australia

Local Arrangement Chair

Simrandeep Singh   IIT Ropar, India

Contents

Fine Tuned Single Shot Detector
for Finding Disease Patches in Leaves

Divyansh Thakur[1]([✉]), Jaspal Kaur Saini[1], and Srikant Srinivasan[2]

[1] School of Computing, Indian Institute of Information Technology Una,
Una, Himachal Pradesh, India
{divyansh,jaspalkaursaini}@iiitu.ac.in
[2] Plaksha University, Chandigarh, India
srikant.srinivasan@plaksha.edu.in

Abstract. Plant disease detection is an important aspect of modern
agriculture that is crucial for ensuring crop productivity and quality.
Sweet lime is an important citrus fruit, and its leaves are susceptible
to a range of diseases that can significantly impact its yield. However,
the lack of publicly available data on sweet lime diseases has made the
development of effective detection systems challenging. To address this
problem, the authors of this paper developed their own dataset of sweet
lime leaves containing 4000 images. The dataset was carefully curated to
ensure diversity and accuracy, and it was used to train a fine-tuned customised single-shot detector (SSD) model. The SSD is a popular object
detection algorithm that is known for its speed and accuracy, making it
well-suited for this task. The results of the evaluation showed that the
proposed approach achieved impressive performance metrics. The model
had an accuracy of 99%, which indicates that it was able to correctly
identify the presence or absence of diseases in the sweet lime leaves with
high confidence. The mean intersection over union (mIoU) of 97% is a
measure of how well the model was able to detect the boundaries of the
diseased areas, indicating that it was able to accurately localize the diseases. The model's inference time of 16ms and frames per second (FPS)
of 60 demonstrate that it is fast enough to be used in real-time applications. Finally, the mean average precision (mAP) of 0.97 is a measure
of how well the model was able to rank the detected diseases by their
severity, demonstrating its effectiveness in prioritizing diseases for further
action. These findings have important practical implications for agricultural management. The proposed approach could be used to monitor
sweet lime orchards for diseases and to identify the specific types of diseases present. This information could be used to make targeted interventions, such as applying fungicides or removing infected leaves, to prevent
the spread of diseases and ensure the health and productivity of the crop.
Furthermore, the methodology used in this study could be adapted for
use with other crops and diseases, expanding its potential impact in the
field of agriculture.

Keywords: Sweet lime dataset · Plant disease detection · Single Shot
Detector

M. K. Saini et al. (Eds.): ICA 2023, CCIS 1866, pp. 1–14, 2023.
https://doi.org/10.1007/978-3-031-43605-5_1

1   Introduction

The integration of technology in the agriculture sector has revolutionized the way we produce, process, and distribute food globally. As the world's population continues to grow, the demand for food is increasing, and farmers face the challenge of meeting these demands while minimizing their environmental footprint. To achieve sustainable and efficient agriculture practices, technology has become an indispensable tool. Advancements in precision agriculture, robotics, artificial intelligence, and data analytic have transformed the agriculture industry, enabling farmers to increase crop yields, reduce production costs, and improve food quality. By utilizing AI algorithms,drones, sensors, and GPS systems, farmers can gather accurate information about soil quality, moisture levels, and plant growth rates, enabling them to make informed decisions about when to plant, water, and harvest crops [1].

The use of Artificial Intelligence (AI) in the agriculture sector has revolutionized the way farming is practiced. With the advent of AI technology, farmers can now analyze vast amounts of data on crops, weather patterns, soil quality, and pest infestations to make informed decisions that enhance crop yield and reduce wastage. AI has enabled the creation of smart farms, where technology is integrated with farming practices to optimize resources and improve efficiency. This has led to increased productivity, reduced costs, and improved sustainability in the agriculture sector [2].

Detecting disease patches in plant leaves is of utmost importance in ensuring healthy plant growth and maximizing crop yields. Disease patches can indicate the presence of pathogens, such as bacteria, fungi, or viruses, which can cause severe damage to plants and result in substantial crop losses. Identifying these patches early on can help prevent the spread of the disease, allowing farmers to take proactive measures, such as targeted treatment, to mitigate the damage. Disease patch detection also plays a critical role in ensuring food safety, as contaminated crops can pose significant health risks to consumers. By using advanced technologies, such as deep learning, to detect disease patches, farmers can make more informed decisions about crop management, leading to better yields, improved crop quality, and increased profitability. Deep learning [3], a subset of artificial intelligence, has been gaining popularity in various industries, including agriculture. With the increasing global population and the need to produce more food, farmers are turning to technology to enhance their yields and optimize resource utilization. Deep learning, which enables computers to learn from large amounts of data, has the potential to revolutionize the agriculture sector by providing valuable insights into crop growth, soil quality, and weather patterns. The application of deep learning in agriculture involves the use of advanced algorithms and techniques to analyze data from various sources, such as drones, satellites, and sensors [4]. By processing this data, deep learning models can help farmers make informed decisions about irrigation, fertilization, and pest control, resulting in improved crop yields and resource efficiency.

Deep learning has become increasingly important for disease patch detection in plant leaves, as it enables more accurate and efficient identification of diseased plants. By analyzing large amounts of data, deep learning algorithms can identify subtle patterns and differences in plant images that are indicative of disease patches, allowing farmers to take proactive measures to mitigate crop damage. The use of deep learning in disease patch detection has several advantages over traditional methods. For one, it allows for a more objective and automated approach, reducing the need for manual inspection of plant leaves. This can save farmers a significant amount of time and resources, allowing them to focus on other critical aspects of crop management. Additionally, deep learning models can process vast amounts of data quickly, enabling farmers to identify disease patches early and take appropriate action to prevent crop losses. In this particular undertaking, we have employed a finely-tuned single shot detector algorithm to accurately detect disease patches in our self-curated data set comprising of sweet lime leaves. The utilization of fine-tuning techniques has led to remarkably good results, exhibiting a high degree of precision in identifying the disease patches.

2 Work Flow

This work is started by creating a data set of sweet lime leaves which contained images of healthy, infected, and decayed leaves. The data set was cleaned to ensure that the data was of high quality and ready for further analysis. Image annotation was done to label the disease patches in the images. After the data was labeled, a custom fine tuned single shot detector (SSD) was trained and tested using both labelled and unlabeled data. Performance parameter results were obtained to evaluate the effectiveness of the model. Finally, the trained SSD was used to detect the disease patches on the sweet lime leaves, successfully completing the work.

3 Related Work

In this particular section, we delve into a comprehensive analysis of existing literature on plant disease detection. Our focus is on exploring the work of various researchers in this field, investigating the methodologies they use, the results they obtain, and their contributions towards advancing plant disease detection technologies. Our aim is to provide a comprehensive overview of the current state of the art in plant disease detection, highlight the significant contributions made by researchers in this area, and pave the way for further research and innovation. The work of Chowdhury et al. [5] aimed to detect diseases affecting tomato plants in fields or greenhouses using deep learning. The algorithm's goal was to run in real-time on a robot to detect plant diseases while moving around the field or greenhouse, or on sensors in fabricated greenhouses to capture close-up photographs of the plants. Durmug et al. [6] highlighted the importance of plants as a primary source of energy production for humans due to their nutritious and medicinal values. However, plant diseases caused significant damage to crop production and economic market value. Therefore, various machine learning (ML) and deep learning (DL) methods were developed and examined by researchers to detect plant diseases. Mahum et al. [7] addressed the challenge of minimizing crop damage for food security by timely detection of diseases in plants. The researchers investigated the use of segmented image data to train convolutional neural network (CNN) models for disease detection. Mohameth et al. [8] proposed the use of computer vision and artificial intelligence (AI) for early detection of plant diseases to reduce the adverse effects of diseases and overcome the limitations of continuous human monitoring. Sharma et al. [9] investigated the use of deep learning and transfer learning techniques for detecting crop diseases, specifically on leaves. They proposed a "smartphone-assisted disease diagnosis" approach using the Plant Village Dataset. Sujatha et al. [10] presented an improved deep learning algorithm for timely detection and classification of potato leaf diseases. The algorithm used a pre-trained Efficient DenseNet model with a reweighted cross-entropy loss function to minimize overfitting during training. Finally, Roy et al. [11] proposed a novel deep learning-based object detection model for accurate detection of fine-grained, multi-scale early plant diseases, specifically in multi-class apple plant disease detection in real orchard environments. In their research work, Vallabhajosyula, et al. [12] proposed an automatic plant disease detection technique using deep ensemble neural networks (DENN). They addressed the challenge of early plant disease detection by fine-tuning pre-trained models through transfer learning and employing data augmentation techniques to overcome overfitting. The proposed approach was evaluated on the publicly available plant village dataset, and the results showed that DENN outperformed state-of-the-art pre-trained models. Another study by Tiwari et al. [13] proposed a deep-learning-based approach for automatic plant disease detection and classification using leaf images captured at various resolutions. They used the Dense Convolutional Neural Network (DCNN) architecture to train on a large plant leaves image dataset from multiple countries and obtained high cross-validation and test accuracy. The proposed method has real-time performance and can potentially save time and resources in plant disease monitoring. In a study by Islam et al. [14], an automated detection approach for paddy leaf disease detection was developed using deep learning Convolutional Neural Network (CNN) models. They focused on four common diseases of paddy leaves in Bangladesh and compared the performance of four different CNN models. The Inception-ResNet-V2 model achieved the highest accuracy of 92.68

4   Data Processing

This section provides critical information about the dataset and the process of data cleaning, which is a crucial aspect of any deep learning model. High-quality data is an indispensable element for achieving accurate and reliable results from a deep learning model. Therefore, Therefore, in this section, we will delve into the details of data set and data cleaning and its significance in ensuring the robustness and accuracy of deep learning models.

4.1 Description of Data Set

The dataset comprised of 4000 images of sweet lime leaves, which were self-clicked by the authors. The images were captured using high-quality cameras and under proper sunlight conditions to ensure good image quality. Among the 4000 images, 1500 were of healthy sweet lime leaves, and 1500 were of diseased leaves, while the remaining 1000 images were of decayed leaves. The images of diseased leaves were captured to encompass a range of common diseases that affect sweet lime leaves. With the help of this dataset, we trained our model to recognize healthy, diseased, and decayed sweet lime leaves with a high degree of accuracy. The high-quality images in the dataset could prove to be beneficial in developing algorithms that can function in real-world situations, such as on a farm, where leaf health analysis is a crucial aspect of plant health management.

4.2 Data Cleaning

In order to prepare the dataset for analysis, data cleaning was conducted through a combination of manual and automated methods. The manual method involved filtering the images to remove any redundant or poor quality ones, which was performed by the researchers. Additionally, a Python script was employed to standardize the size of all images and rename them in a sequential manner from 1 to N, where Nth represents number of last image in the dataset. This automated approach ensured consistency and efficiency in the data preparation process.

5 Material and Methods

This section allows readers to understand how the research was conducted, The purpose of this section is to ensure that the study can be replicated by others and to establish the credibility and validity of the findings. In this section, we provide a detailed description of the materials and methods used in our study to investigate our problem.

5.1 Image Annotation

Image annotation is used to identify objects and their corresponding locations in the image, which enables the deeplearning model to detect them accurately. For this work we annotated the images by using LabelMe software [15]. Annotated images provide information about the location and class of objects within an image. The annotated image sample is shown in Fig. 1.

5.2 TTS Ratio

The train-test ratio for object detection using Single Shot Detection (SSD) depends on the size of the dataset, the complexity of the task, and the amount of available data. The size of the training set is usually larger than the testing set. However, the split can vary depending on the task and dataset. For this work we used train-test ratio of 80:20.

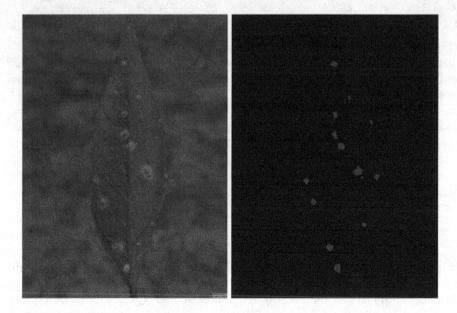

Fig. 1. Annotated images

5.3   GPU

For execution of this work we used NVIDIA V100. The NVIDIA V100 GPU is a high-performance graphics processing unit (GPU) designed for use in data centers and high-performance computing applications. It was first released in 2017 and is based on the Volta architecture, which is known for its ability to handle both traditional graphics workloads and complex artificial intelligence (AI) and deep learning tasks. The V100 GPU is incredibly powerful, with 5,120 CUDA cores and 640 Tensor Cores. It also has a memory bandwidth of up to 900 GB/s and a maximum memory size of 16 GB per GPU. This makes it ideal for demanding applications such as scientific simulations, machine learning, and natural language processing.

5.4   Single Shot Detector

The SSD (Single Shot Detector) architecture is a popular object detection model that is designed to efficiently detect objects of different sizes and aspect ratios in an image. The model is based on a convolutional neural network (CNN) that is pre-trained on a large dataset (such as ImageNet) to learn useful features that can be used for object detection. SSD is a popular object detection algorithm that can detect objects in images and video frames. It was first introduced in 2016 by Wei Liu, Dragomir Anguelov, and other researchers from Google. The main idea behind SSD is to divide the input image into a grid of cells and predict the presence of objects in each cell. Unlike some other object detection algorithms, SSD uses a single neural network to perform both object detection and classification, making it very efficient and fast. The network architecture of SSD consists of a base convolutional neural network, such as VGG or ResNet, followed by a set of convolutional layers that predict the class and location of objects in each cell of the grid. Specifically, these layers predict a set of bounding boxes (i.e., rectangles) around objects and the probability that each bounding box contains an object of a particular class.

In our work we proposed customised and fine tuned the SSD, Table 1 shows the layer architecture of SSD that we use in our work. Here in Table 1, The architecture of the SSD model consists of two main components: a base network and a set of additional layers. The base network is typically a CNN that is pre-trained on a large-scale image classification task. In the case of the SSD model, the base network includes layers up to conv7_1, which is a convolutional layer that outputs a feature map of size $3 \times 3 \times 1024$. The additional layers in the SSD model are designed to capture features at different scales and resolutions. These layers are added on top of the base network and include convolutional layers with different filter sizes and kernel sizes. The detection_conv8_2 layer, for example, is a convolutional layer with 1024 filters and a 3×3 kernel size that takes as input the output from the conv7_1 layer and produces a feature map of size $1 \times 1 \times 1024$. The final layers in the SSD model (including detection_conv9_2, detection_conv10_2, and detection_conv11_2) take as input the output from the detection_conv8_2 layer and apply additional convolutional layers with smaller filter sizes to further refine the predicted bounding boxes and class probabilities. The output of the SSD model is a tensor of shape (batch_size, 1, 1, 4 * num_classes), where batch_size is the number of images in the input batch and num_classes is the number of object classes that the model is trained to detect. This tensor contains the predicted bounding boxes and class probabilities for all the objects in the input image. The asterisk (*) in the output shape (batch_size, 1, 1, 4 * num_classes) represents the multiplication operator. In this case, it indicates that the output tensor has a shape of (batch_size, 1, 1, 4 * num_classes), where the size of the fourth dimension is equal to four times the number of object classes (num_classes). Each bounding box prediction in the output tensor contains four values, corresponding to the coordinates of the top-left and bottom-right corners of the bounding box. Therefore, the size of the fourth dimension in the output tensor is four times the number of object classes to account for the four bounding box coordinate values for each class. Algorithm 1 shows the working of proposed SSD model.

5.5   Performance Parameters

Mean Average Precision (mAP): mAP is the most commonly used performance metric for object detection tasks. It measures the average precision of the model across different levels of recall. In other words, it measures how accurately the model can detect objects of different sizes and aspect ratios. Equation 1 is the mathematical representation of mAP, where N is the number of object classes,

Table 1. Layer architecture of the network

Layer Name	Layer Type	Input Shape	Output Shape
input	Input	(batch_size, 300, 300, 3)	(batch_size, 300, 300, 3)
conv1_1	Conv2D	(batch_size, 300, 300, 3)	(batch_size, 150, 150, 64)
pool1	MaxPooling2D	(batch_size, 150, 150, 64)	(batch_size, 75, 75, 64)
conv2_1	Conv2D	(batch_size, 75, 75, 64)	(batch_size, 38, 38, 128)
pool2	MaxPooling2D	(batch_size, 38, 38, 128)	(batch_size, 19, 19, 128)
conv3_1	Conv2D	(batch_size, 19, 19, 128)	(batch_size, 19, 19, 256)
conv4_1	Conv2D	(batch_size, 19, 19, 256)	(batch_size, 19, 19, 512)
conv5_1	Conv2D	(batch_size, 19, 19, 512)	(batch_size, 19, 19, 512)
conv6_1	Conv2D	(batch_size, 19, 19, 512)	(batch_size, 19, 19, 1024)
conv7_1	Conv2D	(batch_size, 19, 19, 1024)	(batch_size, 10, 10, 1024)
conv8_1	Conv2D	(batch_size, 10, 10, 1024)	(batch_size, 5, 5, 1024)
conv9_1	Conv2D	(batch_size, 5, 5, 1024)	(batch_size, 3, 3, 1024)
conv10_1	Conv2D	(batch_size, 3, 3, 1024)	(batch_size, 1, 1, 1024)
detection_conv4_3	Conv2D	(batch_size, 19, 19, 512)	(batch_size, 19, 19, 4 * num_classes)
detection_fc7	Conv2D	(batch_size, 10, 10, 1024)	(batch_size, 10, 10, 6 * num_classes)
detection_conv6_2	Conv2D	(batch_size, 5, 5, 1024)	(batch_size, 5, 5, 6 * num_classes)
detection_conv7_2	Conv2D	(batch_size, 3, 3, 1024)	(batch_size, 3, 3, 6 * num_classes)
detection_conv8_2	Conv2D	(batch_size, 1, 1, 1024)	(batch_size, 1, 1, 4 * num_classes)
detection_conv9_2	Conv2D	(batch_size, 1, 1, 256)	(batch_size, 1, 1, 4 * num_classes)
detection_conv10_2	Conv2D	(batch_size, 1, 1, 256)	(batch_size, 1, 1, 4 * num_classes)
detection_conv11_2	Conv2D	(batch_size, 1, 1, 256)	(batch_size, 1, 1, 4 * num_classes)

and AP(i) is the average precision for class i.

$$mAP = \frac{1}{N} \sum_{i=1}^{N} AP(i) \tag{1}$$

Frames per Second (FPS): The FPS measures how many frames the model can process per second. Higher FPS values indicate faster processing times, which can be important for real-time applications. Equation 2 is the mathematical equation for evaluating FPS, where t is the time it takes for the model to process one frame.

$$FPS = \frac{1}{t} \tag{2}$$

Inference Time: Inference time measures the time it takes for the model to process a single image. Lower inference times indicate faster processing and can be important for real-time applications.

$$InferenceTime = t \tag{3}$$

Accuracy: The accuracy of an SSD measures how well the model can classify objects. This parameter is particularly important when dealing with complex object classes. Equation 4 is the mathematical equation for geting accuracy,

---

Algorithm 1. Customised SSD Algorithm.

Input image I Object detections

Preprocessing: Resize I to 300×300 and subtract the mean value

Base Network: Pass the preprocessed image through a pre-trained VGG16 network up to 'conv7_1'

Additional Layers: Pass the output of 'conv7_1' through a set of additional convolutional layers ('conv8_1', 'conv8_2', etc.) to generate a set of feature maps at different scales

Default Anchor Boxes: Generate a set of default anchor boxes at each spatial location in the feature maps based on different aspect ratios and scales

Bounding Box Regression: For each anchor box, predict the offset values for the four bounding box coordinates (x, y, width, height) relative to the default anchor box

Objectness Scores: For each anchor box, predict the probability that the box contains an object of interest

Non-Maximum Suppression: Filter out overlapping bounding boxes based on their objectness scores and perform non-maximum suppression to obtain the final set of object detections

---

where TP is the number of true positive detections, TN is the number of true negative detections, FP is the number of false positive detections, and FN is the number of false negative detections.

$$Accuracy = \frac{TP + TN}{TP + TN + FP + FN} \qquad (4)$$

Mean Intersection over Union (mIoU): mIoU is a measure of the overlap between the predicted bounding boxes and the ground truth bounding boxes. It measures how well the model is able to localize the objects in the image. Equation 5 is the mathematical equation of mIoU, where N is the number of objects, and IoU(i) is the intersection over union value for object i.

$$mIoU = \frac{1}{N} \sum_{i=1}^{N} IoU_i \qquad (5)$$

6 Results

Within this section, we present details regarding the hyperparameters employed during the training of our bespoke fine-tuned single shot detector (SSD) model, an evaluation of its performance metrics, as well as the visual output generated by our model. This section serves to demonstrate the robustness and effectiveness of our SSD model in accurately detecting plant diseases.

6.1 Hyper Parameters

Hyperparameter values are essential components of any deep learning model. They are the parameters that are set before the training process begins and can have a significant impact on the performance of the model. In this particular case, the hyperparameter values for a Fine-Tuned Single Shot Detector (FTSSD) for finding disease patches in leaves have been specified in Table 2. The base network used for this model is VGG16, a convolutional neural network (CNN) architecture that is widely used for computer vision tasks. The input shape for the model is (300, 300, 3), which means that the input images are 300 pixels wide and 300 pixels high with 3 color channels (RGB). The batch size for training the model is set to 32. The learning rate for this model is set to 0.001, which determines how much the model adjusts its parameters during training based on the error between predicted and actual values. The optimizer used in this case is stochastic gradient descent (SGD), which is a popular optimization algorithm used in machine learning. The momentum is set to 0.9, which helps the model converge faster by adding a fraction of the previous gradient to the current gradient. The weight decay is set to 0.0005, which is a regularization technique that helps prevent overfitting. The loss function used for this model is MultiBox Loss, which is a combination of localization loss and classification loss. The model has a total of 8 convolutional layers, which helps the model extract useful features from the input image. The total number of parameters in this model is approximately 34 million, which means that the model is quite large and complex.

Table 2. Hyperparameter values for a Fine-Tuned Single Shot Detector (FTSSD) for finding disease patches in leaves.

Hyperparameter	Value
Base Network	VGG16
Input Shape	(300, 300, 3)
Batch Size	32
Learning Rate	0.001
Optimizer	SGD
Momentum	0.9
Weight Decay	0.0005
Loss Function	MultiBox Loss
Num Classes	20
Num Prior Boxes	8732
Convolutional Layers	8
Total Parameters	~34 million

6.2   Performance Parameters Evaluation

The Table 3 provides information on the values obtained for Mean Average Precision (mAP), Frames Per Second (FPS), Inference Time, Accuracy, and Mean Intersection over Union (mIoU). Mean Average Precision (mAP) is a widely used metric for evaluating the performance of object detection algorithms, and a high value of 0.97 indicates that the system is able to accurately detect and classify objects in the input data. The high FPS value of 60 and low inference time of 16 ms indicates that the system is able to process input data quickly and provide results in real-time, which is essential for many real-world applications such as robotics, surveillance, and autonomous driving. The high accuracy value of 99 indicates that the system is able to classify objects with high precision and low false positive rate. Finally, the high value of Mean Intersection over Union (mIoU) of 0.95 indicates that the system is able to accurately localize and segment objects in the input data.

Table 3. Results obtained by different performance parameters.

Performance parameter	Value
Mean Average Precision (mAP)	0.97
Frames Per Second (FPS)	60
Inference Time	16 ms
Accuracy	99
Mean Intersection over Union (mIoU)	0.95

6.3   Visual Output by Model

Table 4 showcases the visual output of the proposed fine-tuned SSD model, wherein the model's accuracy in identifying disease patches is observed to be exceptionally high. The bounding boxes demarcate the specific areas on the sweet lime leaves where the disease patches have been identified by the model. The fine-tuned SSD model demonstrates impressive speed in detecting these disease patches, thereby enhancing its practical applicability in the field of plant disease detection.

Table 4. Visual output by SSD

6.4 Conclusion

In conclusion, this paper presents an innovative approach for sweet lime disease detection that achieved impressive performance metrics using a fine-tuned customised single-shot detector (SSD) model. The proposed method could significantly improve agricultural management by enabling targeted interventions to prevent the spread of diseases and ensure crop health and productivity. Additionally, the methodology used in this study could be extended for use with other crops and diseases, expanding its potential impact in the field of agriculture. Future research could focus on refining the model to further increase its accuracy and performance, as well as developing a user-friendly interface to make it accessible to farmers and agricultural experts. Overall, this study has significant implications for the field of agriculture, highlighting the potential of machine learning techniques to improve crop disease management and increase productivity.

References

1. Thakur, D., Kumar, Y., Kumar, A., Singh, P.K.: Applicability of wireless sensor networks in precision agriculture: a review. Wirel. Pers. Commun. **107**, 471–512 (2019)
2. Eli-Chukwu, N.C.: Applications of artificial intelligence in agriculture: a review. Eng. Technol. Appl. Sci. Res. **9**(4), 4377–4383 (2019)
3. Thakur, D., Saini, J.K., Srinivasan, S.: DeepThink IoT: the strength of deep learning in internet of things. Artif. Intell. Rev. 1–68 (2023). https://doi.org/10.1007/s10462-023-10513-4
4. Thakur, D., Saini, J.K.: The significance of IoT and deep learning in activity recognition. In: Singh, P.K., Wierzchoń, S.T., Pawłowski, W., Kar, A.K., Kumar, Y. (eds.) IoT, Big Data and AI for Improving Quality of Everyday Life: Present and Future Challenges. Studies in Computational Intelligence, vol. 1104, pp. 311–329. Springer, Cham (2023). https://doi.org/10.1007/978-3-031-35783-1_18
5. Durmuş, H., Güneş, E.Q., Kırcı, M.: Disease detection on the leaves of the tomato plants by using deep learning. In: 2017 6th International Conference on Agro-Geoinformatics, pp. 1–5 (2017)
6. Sujatha, R., Chatterjee, J.M., Jhanjhi, N.Z., Brohi, S.N.: Performance of deep learning vs machine learning in plant leaf disease detection. Microprocess. Microsyst. **80**, 103615 (2021)
7. Sharma, P., Berwal, Y.P.S., Ghai, W.: Performance analysis of deep learning cnn models for disease detection in plants using image segmentation. Inf. Process. Agric. **7**(4), 566–574 (2020)
8. Chowdhury, M.E.H., et al.: Automatic and reliable leaf disease detection using deep learning techniques. AgriEngineering **3**(2), 294–312 (2021)
9. Mohameth, F., Bingcai, C., Sada, K.A.: Plant disease detection with deep learning and feature extraction using plant village. J. Comput. Commun. **8**(6), 10–22 (2020)
10. Mahum, R., et al.: A novel framework for potato leaf disease detection using an efficient deep learning model. Human Ecol. Risk Assess. Int. J. **29**(2), 303–326 (2023)
11. Roy, A.M., Bhaduri, J.: A deep learning enabled multi-class plant disease detection model based on computer vision. AI **2**(3), 413–428 (2021)

12. Vallabhajosyula, S., Sistla, V., Kolli, V.K.K.: Transfer learning-based deep ensemble neural network for plant leaf disease detection. J. Plant Dis. Prot. **129**(3), 545–558 (2022)
13. Tiwari, V., Joshi, R.C., Dutta, M.K.: Dense convolutional neural networks based multiclass plant disease detection and classification using leaf images. Ecol. Inf. **63**, 101289 (2021)
14. Islam, M.A., Shuvo, M.N.R., Shamsojjaman, M., Hasan, S., Hossain, M.S., Khatun, T.: An automated convolutional neural network based approach for paddy leaf disease detection. Int. J. Adv. Comput. Sci. Appl. **12**(1), 1–9 (2021)
15. Labelme. https://anaconda.org/conda-forge/labelme

Empirical Analysis and Evaluation of Factors Influencing Adoption of AI-Based Automation Solutions for Sustainable Agriculture

Amit Sood[1]([✉]), Amit Kumar Bhardwaj[1], and Rajendra Kumar Sharma[2]

[1] LM Thapar School of Management, Thapar Institute of Engineering and Technology, Patiala, Punjab, India
asood_phd19@thapar.edu
[2] Thapar Institute of Engineering and Technology, Patiala, Punjab, India

Abstract. Rising food demand emphasizes need to modernize ways of agricultural production. The use of digital technologies in agriculture provide a large avenue for increasing production efficiency with limited resources. Artificial Intelligence (AI) uses agricultural data and generates insights for farmers and facilitators for making better decisions during crop lifecycle management resulting in increased productivity. Despite a large number of benefits and government initiatives, the adoption level of AI-based automation solutions in agriculture is quite low which motivates the present study to focus on recognizing factors influencing adoption of AI-based solutions in agriculture. Based on an integrated framework developed on three eminent theories from Information Systems, this study uses of survey data of farmers and facilitators from Northern India, and further examines the interaction of independent variables and validates the proposed framework using factor analysis and regression analysis. The results show that user expectations, information availability, users' involvement in engagement activities and compatibility of technology solution are key factors influencing intent to adopt AI-based solutions in agriculture. The given framework significantly explains the adoption intention and hence enhances understanding of researchers and practitioners to increase adoption of AI-based solutions for sustainable agriculture.

Keywords: Artificial intelligence · sustainable Agriculture · adoption · exploratory factor analysis · regression analysis

1 Introduction

Agriculture has long history of development associated with human civilization and technology. The nature of agricultural activities has changed from labor-intensive to machine-driven with industrial revolution. In modern age, the digital technologies, such as Big Data, Cloud Computing, Artificial Intelligence (AI),

M. K. Saini et al. (Eds.): ICA 2023, CCIS 1866, pp. 15–27, 2023.
https://doi.org/10.1007/978-3-031-43605-5_2

Machine Learning, Internet of Things (IoT) etc. [1], triggered a new era of agriculture which is named as "Agriculture 4.0" [2]. The focus of researchers to create sustainable methods of enhancing agricultural yield using new digital technologies has increased across globe. [3]. With objective of increasing agricultural production level, solutions based on these technologies use limited resources to meet rising demand in line with Sustainable Development Goal (SDG) of United Nations (UN) to reduce hunger level to zero globally by 2030.

Artificial Intelligence enable farmers at different stages in agriculture, for example, an appropriate time to sow by analyzing weather conditions, to identify crop diseases from image analysis, indicating appropriate time to irrigate to maintain required moisture levels and many other useful insights. In the background, it uses sensors, unmanned aerial vehicles (UAVs), cameras etc. to collect inputs from field in terms of location, temperature, humidity, crop images etc. to send these over network for storage using technologies such as Cloud and Big Data etc. In next stage, the collected data is analyzed using Artificial Intelligence, Machine Learning etc. to provide useful insights regarding crop health, weather pre-dictions, appropriate sowing time and alerts based on regular monitoring. These characteristics of automation and artificial intelligence based solutions are presented in Fig. 1. It is also illustrated that such insights are not available in traditional work practices by farmers which were highly dependent on human intellect and physical observations.

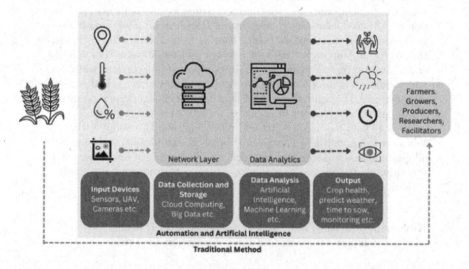

Fig. 1. Technology impact on users with automation and artificial intelligence.

In the field of agriculture, although AI has high potential for exploration by researchers and solution developers to address the concerns related to agricultural operations and production efficiency [1], still there are several challenges which act as deterrents to its adoption in agriculture [2–4]. The low rate of adoption is delaying the achievement of sustainable development goal across many developed and developing countries. With this background, we focus on identifying key factors which influence adoption of AI-based solution. The present study involves two groups - (a) farmers who are the end users of the technology solution and (b) facilitators who act as enablers in the process of implementing new technology solutions which include agricultural universities, government agencies, private organisations etc. The remainder of the paper covers research methodology in Sect. 2 which includes literature review and methodology implemented. Section 3 presents data analysis for the data sets of farmers and facilitators. The results are discussed in Sect. 4 and finally Sect. 5 presents conclusion of the study.

2   Research Methodology

This section discusses the methodology adopted to empirically analyze and evaluate factors influencing adoption of AI-based automation solutions for sustainable agriculture. The flow of activities carried out for present study are presented in Fig. 2.

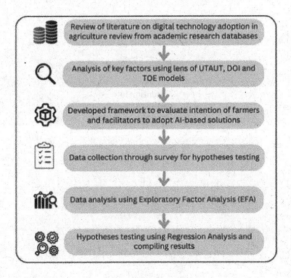

Fig. 2. Flow diagram of the research work carried out in present study.

2.1   Literature Review

Review of literature is carried out using SCOPUS, Web of Science and ACM academic research databases and developed framework based on three well-known theories from Information Systems - Unified Theory of Acceptance and Use of Technology (UTAUT) [5], Technology-Organization-Environment (TOE) Framework [6] and Diffusion of Innovation (DoI) [7] in our previous work [8]. While

UTAUT model evaluates adoption of a new technology in an organized manner, DoI explains how innovation spreads in an organization and TOE presents a combined perspective of Technology, Organization and Environmental factors impacting technology adoption.

The proposed framework provides a total of fifteen independent variables selected from literature that have shown impact on the intention to adopt digital technology solution in agriculture. For improved understanding and analysis, the selected variables are grouped under categories - individual characteristics, environmental factors, structural factors, technology factors and demographic factors. The interaction of grouped factors with dependent variable is visually represented in Fig. 3. Based on proposed framework we developed questionnaire for collecting inputs from farmers and facilitators for analysis using Exploratory Factor Analysis (EFA) and Regression Analysis.

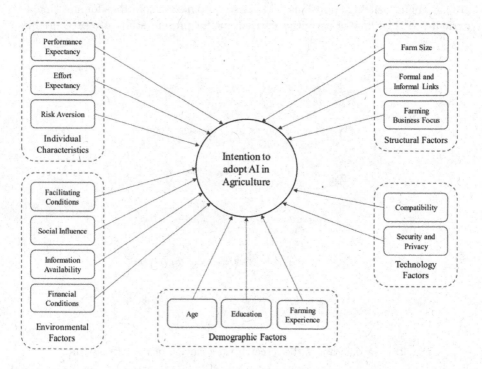

Fig. 3. Literature-based integrated framework for research.

2.2   Methodology

With objective of recognizing significant factors that influence intent of users to adopt AI-based solutions in agriculture, we develop a questionnaire to collect inputs using the lens of proposed framework [8] as explained below:

Instrument Development

To comprehensively understand the adoption factors, this study is divided into two parts: (a) studying adoption intent of farmers who are end users of the solution; (b) evaluating adoption preference of facilitators who enable the implementation of solution in field. The developed questionnaire with 45 questions based on 5-point LIKERT scale evaluates the intention of farmers and facilitators across five dimensions - Individual Characteristics, Environmental Factors, Structural Factors, Technology Factors and Demographic Factors based on the fifteen factors of proposed framework.

Data Collection

The data is collected from the states of Punjab and Haryana which are major contributors of agricultural production in India. The process of data collection uses two separate set of questionnaires for farmers and facilitator groups in three languages including English, Hindi and Punjabi. Inputs are collected from 495 farmers contacted through government offices, local contacts, agricultural university, extensions and co-operatives. The present analysis utilizes 486 responses after excluding nine incomplete responses. Similarly, for facilitator group, data is collected with snowball sampling method. The survey inputs from 198 respondents are used in analysis out of total 205 respondents excluding seven incomplete responses.

3 Data Analysis

The data obtained through survey is used for analysis in three stages: (a) descriptive statistics, (b) Exploratory Factor Analysis (EFA) and (c) Regression Analysis as explained in Sect. 2. The techniques of EFA and Regression Analysis are utilized separately for datasets of farmers and facilitator groups.

3.1 Characteristics of Respondents Included in Study

Descriptive statistics is utilized to analyze the characteristics of respondents included in the present study as summarized in Table 1. Representation of farmers and facilitators from Punjab state is 61% and 57% respectively and remaining from Haryana state. Majority of the farmers (24%) have farming experience between 11 to 20 years and facilitators (30%) with 5–10 years of experience. Largely the respondents from both groups indicate small and marginal farm size i.e. below 2 ha. A very high percentage of smartphone usage, 86% farmers and 92% facilitators is also indicated which is a positive indicator for increasing penetration of digital technology through information dissemination using mobile phone.

Table 1. Summary of characteristics of respondents included in the study

Category	Farmers		Facilitators	
	Frequency	Percent (%)	Frequency	Percent (%)
State				
Punjab	295	61%	113	57%
Haryana	191	39%	85	43%
Gender				
Female	121	25%	44	22%
Male	365	75%	174	88%
Age				
Below 20 years	50	10%	–	0%
21–29 years	129	27%	76	38%
30–39 years	109	22%	55	28%
40–49 years	120	25%	44	22%
Above 50 years	78	16%	23	12%
Experience				
Below 4 years	110	23%	24	12%
5–10 years	106	22%	59	30%
11–20 years	115	24%	51	26%
21–30 years	78	16%	36	18%
More than 30 years	77	16%	28	14%
Education				
No education	12	2%	–	0%
Primary school	43	9%	–	0%
High school	134	28%	2	1%
Senior Secondary	109	22%	46	23%
Graduate or above	188	39%	150	76%
Farm Size				
Marginal (Below 1 ha i.e.<2.47 acres or killas)	85	17%	65	33%
Small (1 to 2 ha i.e. 2.47 to 4.94 acres or killas)	147	30%	56	28%
Semi-medium (2 to 4 ha i.e. 4.94 to 9.88 acres or killas)	140	29%	37	19%
Medium (4 to 10 ha i.e. 9.88 to 24.71 acres or killas)	77	16%	28	14%
Large (10 ha and above i.e. above 24.71 acres or killas)	37	8%	12	6%
Associated with Organization				
Co-operative	243	50%	69	35%
Extension	17	3%	34	17%
Agricultural University	111	23%	48	24%
Private organization	49	10%	32	16%
Others	66	14%	15	8%
Languages known				
Regional language	258	53%	28	14%
Regional language and English	228	47%	170	86%
Smartphone usage				
No	70	14%	16	8%
Yes	416	86%	182	92%

3.2   Analysis Using Inputs from Farmers

EFA is one of the major statistical techniques used in data analytics and research to condense a large number of variables into a smaller set of uncorrelated factors to maximize the variance of these components [9] as represented by following equation:

$$D_i = \Lambda F_j + e_i \tag{1}$$

where D_i is the observed variable in the study, F_j is the common factor, e_i is error term associated with single observed i^{th} variable and i > j as a number of i variables get condensed into j factors. The results of EFA on dataset of farmers yields five factors with eigen value greater than one and explained variance of 64% using principal component analysis with varimax rotation. The summary of extracted factors which influence the intention of farmers to adopt AI-based solutions in agriculture ($ITA_{farmers}$) is given in Table 2 and are interpreted as below:

- Factor-1 is labelled as 'User Expectations' which includes eight observed variables dealing with a combination of performance expectancy and effort expectancy.
- Factor-2 is interpreted as 'Technology Factors' which covers total seven items related to compatibility, security and privacy while using AI-based solutions in agriculture.
- Factor-3 comprises of four items associated with facilitating conditions and two items with financial conditions, hence labelled as 'Facilitating Conditions'.
- Factor-4 is a combination of items related to information availability and formal and informal structures with which farmers are connected with other communities. This factor is interpreted as 'Information and Involvement'.
- Factor-5 includes the component of influence of social elements on the intention of farmers to adopt and is interpreted as 'Social Influence'.

The revised framework with extracted factors is presented in Fig. 4 showing their influence on adoption intent and hence following hypotheses are framed:

H1: User expectations positively influence the intention to adopt AI-based solutions in agriculture.

H2: Technology factors positively influence the intention to adopt AI-based solutions in agriculture.

H3: Facilitating conditions positively influence the intention to adopt AI-based solutions in agriculture.

H4: Information and involvement positively affects the intention to adopt AI-based solutions in agriculture.

H5: Social influence positively impacts the intention to adopt AI-based solutions in agriculture.

The revised framework is further examined using Regression Analysis to understand the significant factors influencing intent of farmers towards adoption

Table 2. Interpretation of latent variables for dataset of farmers

Factor	Interpreted Factor Name	Sum of Squared Loading	Proportion Variance	Original Constructs	Number of Items (Factor Loading >0.5)
Factor-1	User expectations (PUE)	5.01	0.16	Performance expectancy	4
				Effort expectancy	4
Factor-2	Technology factors (TEC)	4.76	0.15	Compatibility	4
				Security and Privacy	3
Factor-3	Facilitating conditions (FAC)	3.76	0.12	Facilitating conditions	4
				Financial conditions	2
Factor-4	Information and Involvement (INF)	3.65	0.12	Information availability	3
				Formal and Informal Link	1
Factor-5	Social Influence (SIN)	2.77	0.09	Social Influence	4

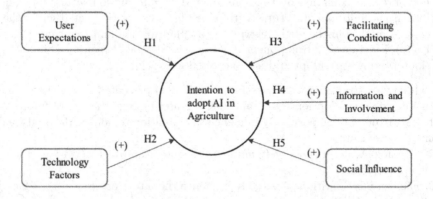

Fig. 4. Revised framework with extracted factors.

of AI-based solutions in agriculture. Regression analysis is widely used statistical technique which is useful in research and business decision making. Using regression analysis, the predictive power of a model can be explained and predictors are evaluated for their relative importance. In the present model for farmers, the regression equation is expressed as below:

$$ITA_{farmers} = \alpha_0 + \alpha_1 A_1 + \alpha_2 A_2 + \alpha_3 A_3 + \alpha_4 A_4 + \alpha_5 A_5 + \varepsilon_{farmers} \quad (2)$$

where α_0 is the intercept, α_1 to α_5 are regression coefficients of predictor variables A_1 to A_5 i.e. PUE, TEC, FAC, INF and SIN, respectively (see Table 2). The farmers' intention to adopt AI-based solutions in agriculture using the proposed model resulted in R^2 of 0.7178 i.e. the set of variables explain about 72% variance in intention to adopt AI-based solutions in agriculture. High value of adjusted R^2 indicates good explanatory power of the model [9]. A visual representation of the results is shown in Fig. 5. The results suggest that all five factors are statistically significant in influencing the intention to AI-based solutions in agriculture. User expectations (PUE), Technology factors (TEC), Facilitating conditions (FAC), Information and Involvement (INF) and Social Influence (SIN) are significant predictors of adopting AI-based solutions with p-values less than 0.001, therefore, all the five hypotheses H1–H5 are supported.

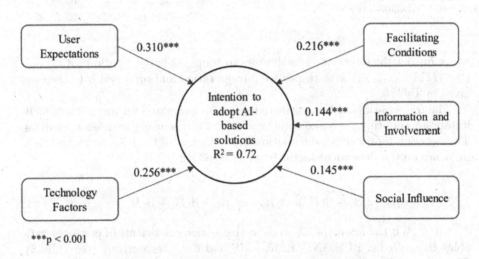

Fig. 5. Summary of results for evaluating intention of farmers.

3.3 Analysis Using Inputs from Facilitators

To understand the perceptions of facilitators, who are enablers in the agricultural system for implementation of the new technology such as agricultural universities, government agencies etc., we adopt similar methodology as used in previous section. Applying EFA on the data set of facilitators, yields five factors with eigen value greater than one which are able to explain the variance of 64% using principal component analysis and varimax rotation. The summary of extracted factors

Table 3. Interpretation of latent variables for dataset of facilitators

Factor	Interpreted Factor Name	Sum of Squared Loading	Proportion Variance	Original Constructs	Number of Items	Hypothesis
Factor-1	User expectations (PUE)	5.41	0.21	Performance expectancy	4	H1 (+)
				Effort expectancy	4	
Factor-2	Information and Involvement (INF)	3.73	0.14	Information availability	4	H2 (+)
				Formal and Informal Link	3	
Factor-3	Perceived Risks (RSK)	2.44	0.09	Risk aversion	4	H3 (−)
Factor-4	Compatibility (CMP)	2.43	0.09	Compatibility	4	H4 (+)
Factor-5	Facilitating conditions (FAC)	1.57	0.06	Facilitating conditions	4	H5 (+)

which impact the intention of facilitators to adopt AI-based solutions in agriculture ($ITA_{facilitators}$) with respective interpretation and proposed hypotheses is given in Table 3.

The framework based on extracted factors is evaluated for intention of facilitators for adopting AI-based solutions in agriculture using regression analysis. The equation representing the relationship of factors in the framework and their influence on the dependent factor is represented as:

$$ITA_{facilitators} = \beta_0 + \beta_1 B_1 + \beta_2 B_2 + \beta_3 B_3 + \beta_4 B_4 + \beta_5 B_5 + \varepsilon_{facilitators} \quad (3)$$

where β_0 is the intercept, β_1 to β_5 are regression coefficients of predictor variables B_1 to B_5 i.e. PUE, INF, RSK, CMP and FAC, respectively (see Table 3). The value of R^2 of the given framework for facilitators is 69% which indicates that the set of factors is able to be explain the variance in the intention of facilitators for adopting AI-based solutions. The results are illustrated in Fig. 6 which indicate that four factors are statistically significant to influence the intention of facilitators to adopt AI-based solutions in agriculture. User expectations (PUE), Information and Involvement (INF) and Compatibility (CMP) are significant predictors of adopting AI-based solutions with p-values less than 0.001 and Perceived risks (RSK) with p-value < 0.01, while Facilitating conditions (FAC) having p-value 0.833 is non-significant. Therefore, four hypotheses H1–H4 are supported. It is important to note that Perceived risks are negatively influencing the intention of facilitators for adopting AI-based solutions.

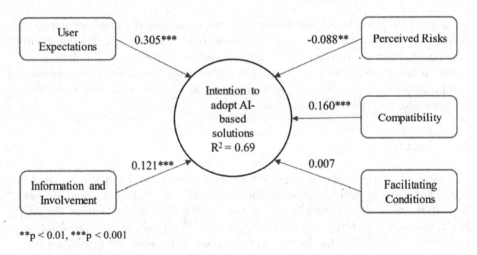

**p < 0.01, ***p < 0.001

Fig. 6. Summary of results for evaluating intention of facilitators.

4 Discussion

The results of this study suggest that User expectations, Technology factors, Facilitating conditions, Information and Involvement and Social Influence the significant factors which drive the intention of farmers to adopt AI-based solutions in agriculture. The level of intention to adopt new technology increases when users relate expected benefits of new technology [10] as indicated in results for the factor User Expectations with highest coefficient ($\beta = 0.310$ for farmers and $\beta = 0.305$ for facilitators). The two dimensions - compatibility, security and privacy represented by technology factors demonstrate fitting of new solution with existing practices, farmer's work style and existing set-up at farm [11,12]. Facilitating conditions play a vital role towards farmers' intention to adopt AI-based solutions in agriculture. The resources in form of experiential trainings and infrastructural aids facilitate farmers to gain knowledge and experience of new technology [13,14]. With such supporting conditions, farmers' trust on the new technology deepens and they get encouraged to start using it [15]. The fourth factor Information and Involvement represents availability of information on AI-based solutions and involvement of farmers in agricultural programs. Farmers learn benefits and features of the new technology using available information, [16,17]. The social engagement by local communities and co-operatives enhances the level of adoption for new technology and practice [13–15] as these project improvement in farmers' work practices. Thus, Social influence also plays an important role towards adoption intent of farmers.

The second part of the study deals with evaluation of perceptions of facilitators towards adopting AI-based solutions in agriculture. The results indicate that User expectations, Information and Involvement, and Compatibility are significant predictors towards adoption intent of facilitators in the similar way as observed for farmers. Additionally, the perceived risks are negatively influencing the adoption intent due to the concerns regarding the new technology in terms of its high cost, efficiency and technical risks. Many of the facilitators are concerned about the initial cost involved in deploying the new technology while others have apprehensions on its performance. Thus, the framework evaluates comprehensive view of factors which influence intent of farmers and facilitators for adopting AI-based solutions in agriculture.

5   Conclusion

The present study evaluates the factors which motivate for adoption of AI-based solutions in agriculture from the perspective of farmers and facilitators. The study uses a framework based on well-known theories to analyze the survey responses from farmers and facilitators with EFA and Regression Analysis techniques. Using EFA the factors from original framework are condensed into five factors which are further analyzed using regression. Our findings indicate that user perceived benefits, information availability, involvement of users and compatibility with existing work practices encourage farmers and facilitators to adopt AI-based solutions. However, perceived risks from the new technology act as deterrent for facilitators towards its adoption. The results of the study can be used by researchers, solution developers and policy-makers to develop improved solutions and devise strategies for higher acceptance of the new technology. While the present study has observed intention of farmers towards adoption of AI-based solutions in Indian context, future research can examine validity of this framework for other emerging economies. Moreover, the present study offers a generalized view of adoption behavior of farmers, while future studies can focus on specific crops or agricultural domains where AI finds large number of applications.

References

1. Klerkx, L., Jakku, E., Labarthe, P.: A review of social science on digital agriculture, smart farming and agriculture 4.0: new contributions and a future research agenda. NJAS-Wageningen J. Life Sci. **90**, 100315 (2019)
2. Sood, A., Sharma, R.K., Bhardwaj, A.K.: Artificial intelligence research in agriculture: a review. Online Inf. Rev. **46**(6), 1054–1075 (2022)
3. Srinivasan, K., Yadav, V.K.: An empirical investigation of barriers to the adoption of smart technologies integrated urban agriculture systems. J. Decis. Syst., 1–35 (2023)
4. Thomas, R.J., O'Hare, G., Coyle, D.: Understanding technology acceptance in smart agriculture: a systematic review of empirical research in crop production. Technol. Forecast. Social Change **189**, 122374 (2023)
5. Venkatesh, V., Morris, M.G., Davis, G.B., Davis, F.D.: User acceptance of information technology: toward a unified view. MIS Q., 425–478 (2003)
6. Tornatzky, L.G., Fleischer, M., Chakrabarti, A.K.: Processes of Technological Innovation. Lexington books (1990)
7. Rogers, E.M.: Diffusion networks. Netw. Knowl. Econ., 130–179 (2003)

8. Sood, A., Bhardwaj, A.K., Sharma, R.K.: Towards sustainable agriculture: key determinants of adopting artificial intelligence in agriculture. J. Decis. Syst., 1–45 (2022)
9. Hair, J.F., Black, W.C., Babin, B.J., Anderson, R.E., Black, W.C., Anderson, R.E.: Multivariate data analysis, eighth. Cengage Learning, EMEA (2019)
10. Mohammad Hossein Ronaghi and Amir Forouharfar: A contextualized study of the usage of the internet of things (iots) in smart farming in a typical middle eastern country within the context of unified theory of acceptance and use of technology model (utaut). Technol. Soc. **63**, 101415 (2020)
11. Drewry, J.L., Shutske, J.M., Trechter, D., Luck, B.D., Pitman, L.: Assessment of digital technology adoption and access barriers among crop, dairy and livestock producers in wisconsin. Comput. Electron. Agric. **165**, 104960 (2019)
12. Balafoutis, A.T., Van Evert, F.K., Fountas, S.: Smart farming technology trends: economic and environmental effects, labor impact, and adoption readiness. Agronomy **10**(5), 743 (2020)
13. Giua, C., Materia, V.C., Camanzi, L.: Smart farming technologies adoption: which factors play a role in the digital transition? Technol. Soc. **68**, 101869 (2022)
14. Michels, M., Bonke, V., Musshoff, O.: Understanding the adoption of smartphone apps in crop protection. Precis. Agric. **21**, 1209–1226 (2020)
15. Heldreth, C., Akrong, D., Holbrook, J., Su, N.M.: What does AI mean for smallholder farmers? a proposal for farmer-centered AI research. Interactions **28**(4), 56–60 (2021)
16. Hermanus Jacobus Smidt and Osden Jokonya: Factors affecting digital technology adoption by small-scale farmers in agriculture value chains (AVCS) in South Africa. Inf. Technol. Dev. **28**(3), 558–584 (2022)
17. Pfeiffer, J., Gabriel, A., Gandorfer, M.: Understanding the public attitudinal acceptance of digital farming technologies: a nationwide survey in Germany. Agric. Human Values **38**(1), 107–128 (2021)

FusedNet Model for Varietal Classification of Rice Seeds

Nitin Tyagi[1]($\boxtimes$)(iD), Yash Khandelwal[2](iD), Pratham Goyal[2](iD), Yash Asati[2](iD),
Balasubramanian Raman[1](iD), Indra Gupta[2](iD), and Neerja Garg[3](iD)

[1] Department of CSE, Indian Institute of Technology Roorkee, Roorkee 247667, India
{nitin_t,bala}@cs.iitr.ac.in
[2] Department of EE, Indian Institute of Technology Roorkee, Roorkee 247667, India
{yash_k,pratham_m,yash_a,indra.gupta}@ee.iitr.ac.in
[3] CSIR-Central Scientific Instruments Organisation, Chandigarh 160030, India
neerjamittal@csio.res.in

Abstract. Rice is one of the most cultivated crops in the world and a primary food source for more than half of the global population. The primary focus of this paper is to implement an ensemble learning model, i.e., FusedNet, that aims to precisely classify 90 different rice seed varieties by utilizing both Red-Green-Blue (RGB) and hyperspectral images (HSI). The FusedNet model comprises two classifiers: first, the support vector machine (SVM) classifier that utilizes spatial and spectral features extracted from the RGB and hyperspectral image data, respectively, and second, the ResNet-50 network (based on Convolutional Neural Network) trained using single seed RGB image data. The model proposed in this study achieved an impressive testing accuracy score of 87.27% and an average F1-score of 86.87%, surpassing the results of the prior investigation conducted on the same publicly available dataset.

Keywords: Hyperspectral Image (HSI) · Convolutional Neural Network (CNN) · Support Vector Machine (SVM) · Ensemble Learning · ResNet-50

1 Introduction

Rice is a widely consumed food crop all over the world. About 50% of the world's global population consumes rice regularly to get nutrients, including carbohydrates, proteins, fats, vitamins, minerals, and dietary fibres. China and India are the top two countries in the world for rice cultivation, with China ranking first and India ranking second in rice production. The quality of rice is linked directly or indirectly to human health. Different rice varieties have varying levels of nutritional value and health benefits. Moreover, the type of rice can also affect the final product's texture, flavor, and aroma.

Y. Khandelwal, P. Goyal, Y. Asati—These authors contributed equally to this work.

M. K. Saini et al. (Eds.): ICA 2023, CCIS 1866, pp. 28–42, 2023.
https://doi.org/10.1007/978-3-031-43605-5_3

Traditionally, the rice varieties were identified manually by observing the morphological traits, color, and texture with the assistance of experienced professional employees. Moreover, High-Performance Liquid Chromatography (HPLC) and Gas Chromatography-Mass Spectrometry are chemical techniques that can be used to identify rice varieties. They are often time-consuming and require costly equipment. Additionally, the aforementioned methods are laborious and lead to the disintegration of the rice samples. Developing a real-time, accurate, and non-destructive automated method for rice variety identification would enable a more efficient allocation of human resources and deliver rapid and consistent outcomes. When used in conjunction with computer vision, hyperspectral imaging [1] shows excellent potential in precisely categorizing rice seeds. The automatic, real-time approaches for classifying rice seeds utilize the following features, including morphological traits, color features, and texture features fetched from the RGB images and spectral traits fetched from the HSI data. Huang et al. [2] utilized a back propagation neural network (BPNN) to classify three paddy seed varieties by extracting spatial features from RGB images. The study reported a classification accuracy of 95.56%. Hong et al. [3] employed a random forest classifier to distinguish six rice varieties from northern Vietnam, utilizing morphological traits such as color, shape, and texture extracted from RGB images. The model demonstrated a discriminative accuracy of 90.54%. Kuo et al. [4] applied a sparse coding method to classify 30 rice varieties by utilizing their morphological features fetched from the RGB images and obtained the highest accuracy of 89.1%. Jin et al. [5] achieved an impressive accuracy of 95% in classifying ten rice varieties by utilizing a convolutional neural network on hyperspectral images. A detailed literature review was explored by Fabiyi et al. [6] on the classification of seeds.

The present study utilizes the fusion of two classifiers to differentiate 90 rice seed varieties. The first classifier is an SVM-based model trained using nine spatial features (perimeter, area, solidity, eccentricity, aspect ratio, major axis length, minor axis length, extent, and maximum Feret diameter) and spectral features (mean spectra and variance). The second classifier utilizes ResNet-50 architecture (based on CNN) trained using the single seed RGB image dataset. The single seed RGB image dataset was prepared by cropping the individual seeds from the original RGB images. The seeds were cropped based on their bounding boxes to preserve the boundary features. Moreover, each seed image was placed on a 350×350 black background to standardize the image size. Finally, both classifiers were fused through an ensemble approach, which resulted in improved classification accuracy and F1-score. This improvement can be attributed to the fact that the CNN-based Resnet-50 classifier extracted additional features from the RGB images, which were then integrated with the SVM classifier's results.

Paper Contributions: (1) Developed machine learning models and trained them using spatial and spectral features extracted from RGB and hyperspectral images, respectively, and compared their performance to obtain the optimal model (Classifier 1). (2) Prepared a dataset of individual seed images cropped from the original RGB images. (3) Developed a second classifier (Classifier 2) based on ResNet-50 architecture and trained it using the prepared single seed RGB image dataset. (4) Combined the optimal machine learning model (Classifier 1) with the ResNet-50-based model (Classifier 2) using weighted average approach and classified rice seeds according to their variety. Further, evaluated the proposed model using various performance metrics.

The structure of the paper is as follows: A concise introduction to the dataset and its preprocessing is presented in Sect. 2, while Sect. 3 elaborates on the modeling techniques employed. Section 4 offers a detailed analysis of the experiments conducted and their outcomes, and the study is concluded in Sect. 5 with a brief summary of future research potential.

2   Materials and Methods

2.1   Original Dataset Description

The dataset under study consists of ninety different varieties of rice seeds, contributed by the National Centre of Protection of New Varieties and Good Plants (NCPNVGP), Vietnam. These varieties were carefully chosen as they are among the most commonly cultivated and consumed rice seed types. Skilled and trained personnel manually ensured that each seed of the dataset belonged to the mentioned variety. A seed sample of 96 seeds for each of the 90 varieties was analyzed in this study. The seeds were then divided into two bundles, with each containing 48 seeds. These 48 seeds were arranged in an 8 × 6 matrix form. Consequently, the data collected in this study corresponds to a total of 8640 seeds (90 varieties × 96 seeds per variety).

The image data was captured using two systems: a detailed RGB camera and an HSI system. The RGB and HSI systems captured data simultaneously for each bundle. The RGB image captured has a high resolution of 4896 × 3264 pixels. The spectral image comprised reflectance values corresponding to wavelengths ranging from 385 nm to 1000 nm. As a result, the hyperspectral data cube consisted of reflectance values of 256 spectral bands.

Dark and white references were also used in this study. For capturing the dark reference, a shutter was used, while a white spectralon surface was used to capture the white reference data. These references were used for the normalization of spectral data. The dataset is described in more detail in [6]. It is publicly available at [7].

The original raw dataset has noise and illumination defects that need to be preprocessed. The RGB and hyperspectral images were processed separately due to their different features of interest. The main features to extract from the high-resolution RGB images are morphological features related to shape and size. Therefore, there is a need to have high-contrast images so as to easily and accurately extract these features. Similarly, the hyperspectral images are required to obtain the reflectance values; therefore, the illumination inconsistencies must be eliminated.

2.2 RGB Image Data Preprocessing and Feature Extraction for Classifier 1

To train Classifier 1, nine morphological traits were fetched from the RGB image dataset. The raw RGB images were cropped to keep only the region with seeds in the frame. The morphological features were extracted using only the red channel out of the three bands (red, green, and blue) since it exhibits the highest contrast with respect to the background.

Next, the images were subjected to a white top-hat transformation to reveal fine details. Then, Otsu's method [8] was applied to threshold the images and convert them to binary images. This method determines a threshold value that separates the foreground objects from the background by maximizing the inter-class variance on the intensity histogram. Subsequently, morphological closing and opening [9] operations were applied to the image. Morphological closing filled all the small holes inside the seeds, making the measurement of features such as area more accurate. Morphological opening and area thresholding were used to eliminate any unwanted noise present in the background. This helped label only the actual seeds and avoid any noise being labeled. In the dataset images, the seeds are arranged in an 8 × 6 grid. To label them, the centroids of the seeds were sorted in a top-to-bottom order and then grouped into buckets of size 6. Within each bucket, the centroids of the seeds were sorted in left-to-right order. This labeling scheme ensured that the seeds were labeled based on their row position first and then in a left-to-right manner within each row, thus establishing a consistent and organized labeling convention as described in Fig. 1.

After processing, the images were utilized to extract nine spatial features, namely perimeter, area, major-axis length[1], minor-axis length(see footnote 1), eccentricity(see footnote 1), aspect ratio(see footnote 1), solidity[2], extent[3], and maximum Feret diameter[4].

2.3 Hyperspectral Data Preprocessing and Feature Extraction for Classifier 1

The mean spectra and variance of the reflectance values of each seed were fetched from the hyperspectral images and fed as input to train Classifier 1. For this, the hyperspectral images need to be preprocessed. The first step in this process involved normalizing the image to correct for any distortions caused by uneven illumination. This was achieved using the standard procedure that has been used in previous works [6,21]. The next step was to create a binary mask to isolate the region of interest, i.e., the seed pixels. This was done by subtracting the

---

[1] These features were extracted by considering the seed region as an ellipse.

[2] Solidity is computed by dividing the shape area by the area of its convex hull.

[3] Extent is computed by dividing the shape area by the area of the smallest rectangle enclosing the shape.

[4] Maximum Feret diameter is defined as the longest distance between two parallel lines restricting the shape.

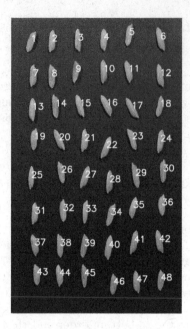

Fig. 1. labeling Scheme

0th band from the 100th band to enhance the contrast of the image and then applying the mean threshold method [10] to create a binary mask. This mask was then subjected to Morphological Opening and Closing transformations, as well as area thresholding to remove any background noise. The resultant binary mask was then used to isolate regions of interest by multiplying it with each band of the normalized hyperspectral image. Next, the seed regions were labeled using the same labeling scheme as described in Sect. 2.2. Following this, the mean and variance of the reflectance values were computed for each of the 256 bands, resulting in a total of 512 hyperspectral features that were utilized in the study.

2.4 Single Seed RGB Image Dataset Preparation for Classifier 2

In the present study, a single seed RGB image dataset is prepared from the original RGB images for training Classifier 2. An overview of the dataset preparation is shown in Fig. 2. To prepare the dataset, binary images were generated from the original RGB images and labeled using the same procedure described in Sect. 2.2. These binary images were then utilized to create a bounding box for each seed. Next, individual seeds were cropped from the original RGB images based on their respective bounding boxes. This approach ensured that the seed boundary region information was retained and incorporated by the classifier. The bounding boxes were of different dimensions for different seeds; therefore, to ensure uniformity in image size for input to the CNN, a black background was placed behind the seed images. This background had a resolution of 350×350 pixels, chosen to fit all the seeds within the image frame.

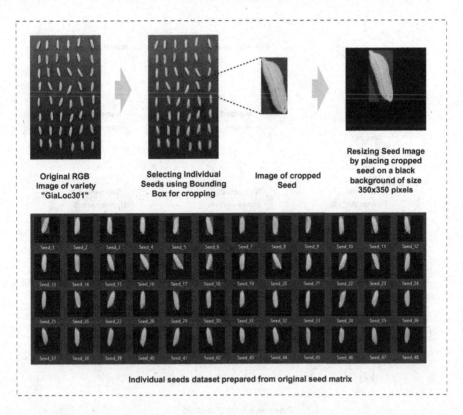

Fig. 2. An overview of single seed RGB image dataset preparation

The preprocessing steps are described in Sect. 2.2, 2.3 and 2.4 are summarised in Fig. 3.

3   Modeling Methods

3.1   Support Vector Machine

The SVM algorithm works by finding the most suitable hyperplane that separates various data classes while maximizing the margin, which refers to the amount of separation between the hyperplane and the closest data points from each class [11]. The SVM algorithm is capable of converting the input data into a higher-dimensional space through the utilization of a kernel function, thereby facilitating the linear separation of the data. Different kernels and hyperparameters were tuned using a grid search technique with 5-fold cross-validation on the training set. Finally, the 'rbf' kernel with penalty term 'C' as 1 and 'γ' as 0.001 gave the best results. The mathematical equation of the radial basis function (RBF) kernel is:

$$K(y_a, y_b) = e^{-\gamma ||y_a - y_b||^2}$$

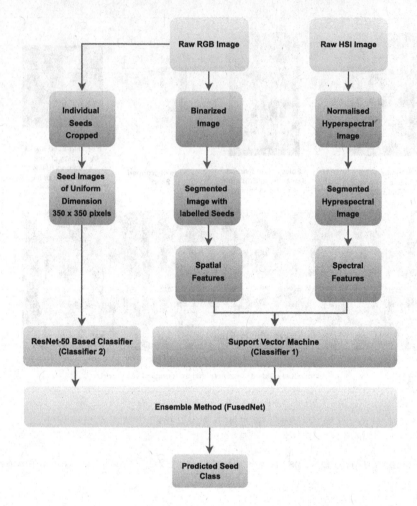

Fig. 3. Schematic representation of the proposed methodology

Here, K is the RBF kernel function, y_a and y_b are data points, γ is a hyper-parameter that governs the width of the RBF kernel, and $||.||$ represents the distance (Euclidean) between y_a and y_b.

3.2 XGBoost

Extreme Gradient Boosting (XGBoost) [12] uses a boosting approach where multiple weak models, usually decision trees, are combined to form a strong model. XGBoost differs from traditional gradient boosting algorithms by incorporating regularization techniques and parallel computing, resulting in improved accuracy and faster training times.

3.3   Random Forest

Random Forest algorithm constructs multiple decision trees using different subsets of the training data and integrates their outputs to produce a final prediction [13]. During the construction of each decision tree, a random set of features is selected, and the criterion for splitting is determined based on the degree of impurity reduction in the target variable.

3.4   Convolutional Neural Network (CNN) and ResNet-50

CNNs [14] use a special type of neuron called a convolutional neuron that applies a mathematical operation known as a convolution on the input image to extract vital features, including edges, corners, and textures. Multiple convolutional layers are stacked together to form a deep neural network that can learn increasingly complex and abstract features from the input image. Additionally, CNNs may incorporate pooling layers to decrease the output size and fully connected layers to make predictions based on the extracted features.

ResNet-50. ResNet stands for Residual Network and is a specific type of convolutional neural network introduced by He *et al.* [15]. ResNet-50 is a pre-trained model (pre-trained on ImageNet dataset [16]) which uses 50 layers in its architecture and uses skip connections to deal with the problem of vanishing gradients which is quite common in deep neural networks.

4   Experiments and Results

4.1   Performance Metrics

The metrics [17] used for evaluating the performance of models in this work are accuracy, precision, recall and F1-score.

$$Accuracy = \frac{Number\,of\,Correct\,Predictions}{Number\,of\,Total\,Predictions} \tag{1}$$

$$Precision = \frac{True\,Positives}{True\,Positives + False\,Positives} \tag{2}$$

$$Recall = \frac{True\,Positives}{True\,Positives + False\,Negatives} \tag{3}$$

$$F1-score = 2.\frac{Precision.Recall}{Precision + Recall} \tag{4}$$

The precision, recall and F1-score metrics described in Eq. 2, 3 and 4 are calculated for each class separately, and then the macro-average of these metrics is computed to describe the performance of the model across all classes.

4.2   Experimental Setup

We conducted our research experiments on a system with specific hardware specifications. The CPU used was an Intel Xeon Gold 5120 processor, which has a base clock speed of 2.2 GHz, 14 cores, and 28 threads, featuring a SmartCache of 19.25 MB. The GPU used an Nvidia Quadro P5000 with 16 GB of GDDR5X VRAM, 2560 CUDA cores, and a memory bandwidth of 288 GB/s. The system was running on the Ubuntu 18.04.6 LTS operating system, and the experiments were conducted using Python 3.9.16 and PyTorch 1.12.1.

4.3   Model Development

In this study, two classifiers have been used, namely, Classifier 1 and Classifier 2, which are further combined, giving the final 'FusedNet' model.

Classifier 1. The RGB and Hyperspectral image dataset was preprocessed, as described in Sects. 2.2 and 2.3, resulting in a dataset with 512 spectral features and 9 spatial features, totalling 521 features. The dataset was partitioned into two sets: a training set consisting of 7344 samples and a testing set consisting of 1296 samples with an 85:15 ratio. To eliminate the redundancy in features, dimensionality reduction was performed using Linear Discriminant Analysis (LDA) [18] which reduced the feature count to 89. The input features were normalized by subtracting the mean and scaling to a standard deviation of one.

The Classifier 1 was chosen from the following three machine learning models: Support Vector Classifier, Random Forest (RF) Classifier and XGBoost Classifier. The hyperparameters of these classifiers were tuned using a grid search technique with 5-fold cross-validation [19]. For Support Vector Classifier, the hyperparameters 'C' (penalty parameter), 'γ' (kernel coefficient), and kernel ('rbf', 'poly' and 'sigmoid') were tuned. To optimize the performance of the Random Forest Classifier, various hyperparameters such as 'n_estimators' (the number of trees in the forest), 'max_depth' (the maximum depth of each tree), 'max_features' (the number of features considered for the best split), and the criterion of split were tuned. To optimize the performance of the XGBoost Classifier, the hyperparameters 'max_depth' (the maximum depth of each tree), 'n_estimators' (the number of estimators), and learning rate were tuned. The best-tuned model of each classifier is presented in Table 1. The Support Vector Classifier (with hyperparameters 'C' = 1, 'γ' = 0.001 and 'rbf' kernel) achieved the highest F1-score of 85.96 compared to the other classifiers. Based on these results, the Support Vector Classifier was selected as Classifier 1. To further analyze the performance of the model, spectral and spatial features were separately used as input data, and the Support Vector Classifier (Classifier 1) was trained on each of them. LDA and standardization were applied to the data the same as before. Table 2 presents a comparison of the performance of Classifier 1 on various combinations of image features. The accuracy achieved by solely utilizing spatial features was 24.31%, whereas utilizing solely spectral features resulted in an accuracy of 79.39%. These results suggest that the selected nine spatial features do not fully capture the information contained in RGB images. Convolution Neural Networks (CNN) have been very successful in image classification tasks, and their ability to extract relevant features from images makes them suitable for this problem. Therefore, to complement the Support Vector Classifier, a second classifier based on CNN was trained on the single seed RGB image dataset.

Table 1. Performance of different machine learning classifiers on spatial and spectral feature data

Classifier name	Hyperparameter values	Accuracy	Average F1-Score	Average Precision	Average Recall
XGBoost	$(2, 0.1, 180)^a$	80.71	79.93	80.44	80.53
RF	$(1000, 15, \text{'auto', 'gini'})^b$	86.34	85.51	85.97	86.52
SVM	$\mathbf{(1, 0.001, \text{'rbf'})^c}$	**86.41**	**85.96**	**86.46**	**86.62**

^{a}(max_depth, learning rate, n_estimators)
^{b}(n_estimators, max_depth, max_features
^{c} (C, γ, kernel)criterion)

Table 2. Performance of Support Vector Classifier on different combinations of image features

Data	Accuracy	Avg. F1-Score	Avg. Precision	Avg. Recall
Spatial features only	24.31	23.05	23.32	26.38
Spectral features only	79.39	78.61	79.77	79.37
Spatial + spectral features	86.41	85.96	86.46	86.62

Classifier 2. The CNN based Classifier 2 was trained using the single seed RGB image dataset prepared in Sect. 2.4. The dataset was partitioned into a training set of 7344 images, which accounted for 85% of the dataset, and a testing set of 1296 images, which accounted for the remaining 15%. The random state used for the train-test split was the same for both Classifier 1 and Classifier 2. Same random state implies that the same seeds whose features were used in the training set of Classifier 1 were present as cropped images in the training set of this classifier. A further division of the training set resulted in a training subset consisting of 5875 images, representing 80% of the dataset, and a validation subset consisting of 1469 images, representing the remaining 20%. During the training process, the loss function employed was categorical cross-entropy, and the Adam optimizer was utilized. Different pre-trained CNN models were trained and validated, and the ResNet-50 [20] model was found to outperform the others in terms of validation accuracy. Hyperparameter tuning was performed using the same training and validation sets. The final Classifier 2 model contained three fully connected layers on top of pre-trained ResNet-50 with ReLU activation

and 30% dropout, achieving an accuracy of 31.01%, F1-score of 30.04%, recall of 31.42%, and precision of 32.28%. Our findings indicate that this CNN-based Classifier 2 (trained on RGB image dataset) performs better than Classifier 1 (trained on only spatial features).

FusedNet. The present study culminates with the integration of the two classifiers discussed earlier to create an ensemble classifier. Figure 4 depicts the architecture of the proposed ensemble classifier. In order to find the final predicted label, the predicted probabilities for the 90 seed varieties were obtained using the two classifiers. Subsequently, the weight coefficients (α and $1-\alpha$ as described in Eq. 5) for the two classifiers were determined through a grid search process using 5-fold cross-validation. Finally, a weighted average of the outputs from both classifiers was calculated to obtain the final predicted label. The predicted probability of ith class generated by ensemble-based FusedNet, $P_{FusedNet_i}$ is given by Eq. 5

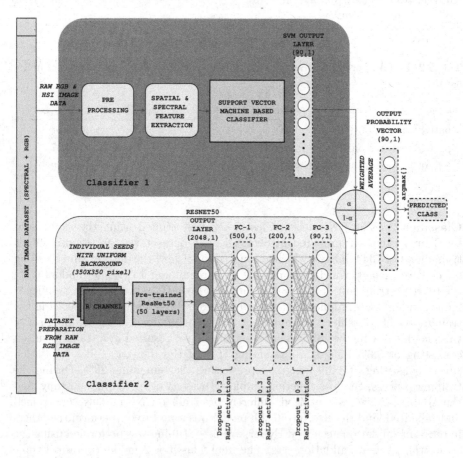

Fig. 4. Schematic representation of proposed FusedNet architecture.

$$P_{FusedNet_i} = \alpha * P_{Clf1_i} + (1 - \alpha) * P_{Clf2_i} \tag{5}$$

Here, P_{Clf1_i}, and P_{Clf2_i} are the probabilities of the ith class generated by Classifier 1 and Classifier 2, respectively. α is the weight parameter corresponding to Classifier 1. The optimal value of the parameter α, say α^* was computed using the following algorithm.

1. Initialize the possible values of $\alpha = [0.00, 0.01, 0.02, 0.03, ..., 0.98, 0.99, 1.00]$.
2. For each possible value of α, say α_j, calculate the $P_{FusedNet_{i,j}}$ for i^{th} class as given in Eq. 5. Then calculate the 5-fold cross-validation accuracy. For each α_j,

$$Accuracy_{avg,j} = \frac{\sum\limits_{all\,folds} \left(\dfrac{\sum\limits_{val\,set} (\arg\max\limits_{i}(P_{FusedNet_{i,j}}) == Y_{seed})}{size(val\,set)} \right)}{number\,of\,folds} \tag{6}$$

where Y_{seed} is the actual class (variety) of the seed and *val set* is the validation set in each fold.
3. Choose that value of α_j as α^* that gives the maximum mean accuracy over all the folds.

$$\alpha^* = \arg\max_{\alpha_j}(Accuracy_{avg,j}) \tag{7}$$

The plot of 5-fold cross-validation accuracy vs α is shown in Fig. 5. The optimal value of α was determined to be 0.90, which resulted in the highest accuracy. This value was then selected as α^*, representing the optimal value for FusedNet's final class prediction. Here, $\alpha^* = 0.9$ implies that Classifier 1 contributes 90% to the prediction, while Classifier 2 contributes 10%. The inclusion of Classifier 2 has played a crucial role in achieving improved accuracy.

The proposed FusedNet Classifier achieved a testing accuracy of 87.27%, an average F1-score of 86.87%, an average precision of 87.36%, and an average recall of 87.53% on the testing dataset.

4.4   Comparison with State-of-the-Art Methods

Table 3 presents a comparison between our proposed approach and the current state-of-the-art methods, all evaluated on the same dataset. The preceding sections discuss the quantitative outcomes, while the qualitative findings, illustrated in Fig. 6, showcase the predicted varieties of seed images.

Table 3. Comparison with State-of-the-art Methods

Author/Model	Avg. F1-Score	Avg. Precision	Avg. Recall
Filipović *et al.* [21]	85.65	86.21	86.00
Fabiyi *et al.* [6]	78.27	79.64	78.80
Proposed FusedNet	**86.87**	**87.36**	**87.53**

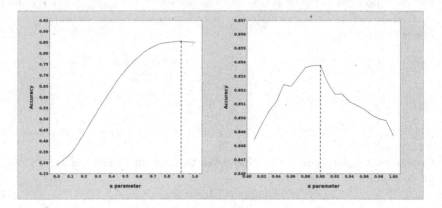

Fig. 5. Plot of cross validation accuracy versus α parameter. The best accuracy was achieved at $\alpha = 0.9$. Figure (b) is a magnified version of Fig. (a) for α in the range [0.80,1.00].

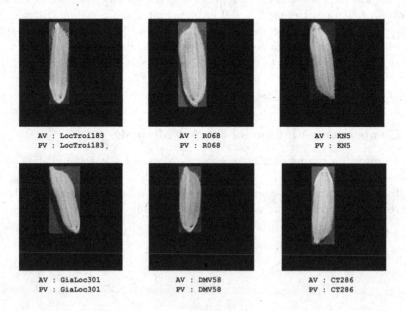

Fig. 6. Predictions for sample data. Here, 'AV': Actual Variety, 'PV': Predicted Variety.

5 Conclusion and Future Work

The proposed FusedNet architecture based on ensemble learning [22] outperformed the machine learning models. The F1-score obtained by the FusedNet architecture is superior to that achieved by existing state-of-the-art methods. Specifically, the CNN-based classifier trained using RGB images outperformed the SVM-based classifier trained using only spatial features, suggesting that CNNs are better equipped to extract features from images that may be missed by traditional methods. Moving forward, the proposed method could be applied to larger datasets, as deep learning models require a significant amount of data to be trained properly. Moreover, the results demonstrate the potential of ensemble learning approaches in improving the accuracy of seed classification models.

Acknowledgement. The Ministry of Education, INDIA, supported the research work with reference grant number: OH-3123200428, and one of the authors, Balasubramanian Raman, acknowledged the financial support from the SERB MATRICS grant under file no. MTR/2022/000187.

References

1. ElMasry, G., Sun, D.-W., Allen, P.: Near-infrared hyperspectral imaging for predicting colour, ph and tenderness of fresh beef. J. Food Eng. **110**(1), 127–140 (2012)
2. Huang, K.-Y., Chien, M.-C.: A novel method of identifying paddy seed varieties. Sensors **17**(4), 809 (2017)
3. Hong, P.T.T., Hai, T.T.T., Hoang, V.T., Hai, V., Nguyen, T.T., et al.: Comparative study on vision based rice seed varieties identification. In: 2015 Seventh International Conference on Knowledge and Systems Engineering (KSE), pp. 377–382. IEEE (2015)
4. Kuo, T.-Y., Chung, C.-L., Chen, S.-Y., Lin, H.-A., Kuo, Y.-F.: Identifying rice grains using image analysis and sparse-representation-based classification. Comput. Electron. Agric. **127**, 716–725 (2016)
5. Jin, B., Zhang, C., Jia, L., Qizhe Tang, L., Gao, G.Z., Qi, H.: Identification of rice seed varieties based on near-infrared hyperspectral imaging technology combined with deep learning. ACS Omega **7**(6), 4735–4749 (2022)
6. Fabiyi, S.D., et al.: Varietal classification of rice seeds using rgb and hyperspectral images. IEEE Access **8**, 22493–22505 (2020)
7. Vu, H., et al.: RGB and VIS/NIR Hyperspectral Imaging Data for 90 Rice Seed Varieties (2019)
8. Talab, A.M.A., Huang, Z., Xi, F., HaiMing, L.: Detection crack in image using otsu method and multiple filtering in image processing techniques. Optik **127**(3), 1030–1033 (2016)
9. Vincent, L.: Morphological area openings and closings for grey-scale images. In: Shape in Picture: Mathematical Description of Shape in Grey-Level Images, pp. 197–208. Springer, Heidelberg (1994). https://doi.org/10.1007/978-3-662-03039-4_13
10. Mardia, K.V., Hainsworth, T.J.: A spatial thresholding method for image segmentation. IEEE Trans. Pattern Anal. Mach. Intell. **10**(6), 919–927 (1988)

11. Hearst, M.A., Dumais, S.T., Osuna, E., Platt, J., Scholkopf, B.: Support vector machines. IEEE Intell. Syst. Appl. **13**(4), 18–28 (1998)
12. Chen, T., et al.: Xgboost: extreme gradient boosting. R Package Vers. 0.4-2 **1**(4), 1–4 (2015)
13. Breiman, L.: Random forests. Mach. Learn. **45**, 5–32 (2001)
14. O'Shea, K., Nash, R.: An introduction to convolutional neural networks. arXiv preprint arXiv:1511.08458 (2015)
15. He, K., Zhang, X., Ren, S., Sun, J.: Deep residual learning for image recognition. In: Proceedings of the IEEE Conference on Computer Vision and Pattern Recognition (CVPR) (2016)
16. Deng, J., Dong, W., Socher, R., Li, L.J., Li, K., Fei-Fei, L.: Imagenet: a large-scale hierarchical image database. In: 2009 IEEE Conference on Computer Vision and Pattern Recognition, pp. 248–255. IEEE (2009)
17. Ferri, C., Hernández-Orallo, J., Modroiu, R.: An experimental comparison of performance measures for classification. Pattern Recogn. Lett. **30**(1), 27–38 (2009)
18. Balakrishnama, S., Ganapathiraju, A.: Linear discriminant analysis-a brief tutorial. Inst. Signal Inf. Process. **18**(1998), 1–8 (1998)
19. Anguita, D., Ghelardoni, L., Ghio, A., Oneto, L., Ridella, S.: The 'k' in k-fold cross validation. In: ESANN, pp. 441–446 (2012)
20. Mukti, I.Z., Biswas, D.: Transfer learning based plant diseases detection using resnet50. In: 2019 4th International Conference on Electrical Information and Communication Technology (EICT), pp. 1–6. IEEE (2019)
21. Filipović, V., Panić, M., Brdar, S., Brkljač, B.: Significance of morphological features in rice variety classification using hyperspectral imaging. In: 2021 12th International Symposium on Image and Signal Processing and Analysis (ISPA), pp. 171–176. IEEE (2021)
22. Zhou, Z.H., Zhou, Z.H.: Ensemble Learning. Springer, Heidelberg (2021)

Fertilizer Recommendation Using Ensemble Filter-Based Feature Selection Approach

M. Sujatha[✉] [iD] and C. D. Jaidhar

Department of Information Technology, National Institute of Technology Karnataka, Surathkal, Mangalore 575025, Karnataka, India
{sujatham.197it002,jaidharcd}@nitk.edu.in

Abstract. Precise application of fertilizer is essential for sustainable agricultural yield. Machine learning-based classifiers are vital in evaluating soil fertility without contaminating the environment. This work uses machine learning-based classifiers such as Classification and Regression Tree, Extra Tree, J48 Decision Tree, Random Forest, REPTree, Naive Bayes, and Support Vector Machine to classify soil fertility. Initially, soil classification was conducted using chemical measurements of 11 soil parameters such as Electrical Conductivity, pH, Organic Carbon, Boron, Copper, Iron, Manganese, Phosphorus, Potassium, Sulphur, and Zinc. The traditional laboratory analysis of soil chemical parameters is time-consuming and expensive. This research work focuses on developing a robust machine learning-based classification approach by employing prominent features recommended by the ensemble filter-based feature selection. To overcome the inconsistency in generating different feature scores, an ensemble filter-based feature selection is devised using three different filter-based feature selection approaches: Information Gain, Gain Ratio, and Relief Feature. Two different datasets are used to evaluate the robustness of the proposed approach. Obtained experimental results demonstrated that the proposed approach with the Random Forest classifier achieved the highest Accuracy of 99.96% and 99.90% for dataset-1 and dataset-2, respectively. The proposed method reduces the inconsistency in feature selection by eliminating a common parameter from both datasets. It minimizes the cost of soil fertility classification by using relevant soil parameters. The classification results are used to provide fertilizer prescriptions.

Keywords: Classifier · Feature Selection · Machine Learning · Soil Fertility · Sustainable Agriculture

1 Introduction

India's development predominantly relies on agricultural development. The increased or decreased soil fertilization deteriorates or accelerates the soil nutrients [13]. The variation in soil nutrients decreases crop yield. Excess fertilization

© The Author(s), under exclusive license to Springer Nature Switzerland AG 2023
M. K. Saini et al. (Eds.): ICA 2023, CCIS 1866, pp. 43–57, 2023.
https://doi.org/10.1007/978-3-031-43605-5_4

causes environmental pollution. It is essential to maintain soil fertility balance to improve agricultural productivity. As a result, it is essential to classify the soil parameters for fertilizer recommendations with environmentally sustainable technologies. Soil classification using machine learning-based classifiers helps sustainable and improved agricultural production [20].

The goal of using a machine learning-based classifier in this work is to classify soil fertility sustainably without affecting the environment. This work uses tree-based classifiers, namely Classification and Regression Tree (CART), Extra Tree, J48 Decision Tree (J48), Random Forest (RF), REPTree (REP), and Random Tree (RT) to classify the soil fertility, and compares them with Naive Bayes (NB) and Support Vector Machine (SVM) classifiers. Soil fertility depends on various chemical parameters such as Electrical Conductivity (EC), pH, Organic Carbon (OC), and nutrition levels existing in the soil. The soil nutrients such as Potassium (K), Nitrogen (N), and Phosphorous (P) are essential in large quantities, whereas Sulphur (S), Boron (B), Copper (Cu), Iron (Fe), Manganese (Mn), and Zinc (Zn) are necessary at lesser quantities for better crop growth and yield [14]. Many former works have employed machine learning-based approaches to estimate soil fertility. A few researchers have recommended fertilizers based on deficiency of soil nutrients N, P, K, or S [2, 17, 20]. In this research, fertilizers are recommended based on soil chemical parameters including *EC, pH, OC, B, Cu, Fe, Mn, K, P, S*, and *Zn*. The deficit of *N* is determined based on the chemical measurement of *pH*.

Traditional laboratory analysis of soil samples is plagued by time constraints and high costs for determining soil nutrients. The filter-based feature selection techniques reduce the computational overhead and complexity of classification by selecting a subset of the most relevant features [15]. It is advantageous to select the relevant soil parameters to minimize the cost of lab analysis. The different filter-based feature selection methods are Correlation, Chi-Squared, Information Gain (InfoG), Gain Ratio (GainR), OneR, and Relief Feature (ReliefF). Correlation determines the linear relationship between two or more features [15]. As the soil parameters are correlated, using a correlation approach to decide on irrelevant features is inappropriate. Chi-squared finds the relevance of features by measuring Chi-squared statistics with respect to the class. It is erratic for the imbalanced dataset [10]. InfoG applies entropy to calculate the weight of the feature. As a result, the class prediction becomes less uncertain. It assesses both relevancy and redundancy of features in predicting the target variable [4]. GainR differs from InfoG in that it incorporates a normalization factor to correct the bias of high-valued features [15]. ReliefF ranks the features according to their ability to separate data points close to each other in the attribute space. It selects a sample instance from the dataset, and its features are compared to its neighbors [5]. The selected features vary for different feature selection techniques and depend on the datasets used [15]. This research proposes a robust soil fertility classification approach using the features selected based on an ensemble filter-based feature selection approach. With the proposed approach, farmers can apply precise amounts of fertilizer with minimal soil nutrient testing.

The prime contributions of this research work are:

- Soil-health data was collected from two different districts of Karnataka (State), India, and labeled as LOW, MEDIUM (MED), or HIGH.
- An ensemble filter-based feature selection approach is proposed by combining three different filter-based feature selection approaches, namely InfoG, GainR, and ReliefF.
- Soil fertility classification approach is proposed based on the features rank given by the ensemble filter-based feature selection approach.
- Based on the classification results, fertilizers are prescribed.

2   Related Work

Many authors have investigated using machine learning-based classifiers to classify soil fertility. Schillaci et al. [19] used Boosted Regression Tree (BRT) to measure OC present in the soil. Sirsat et al. [20] used various machine learning-based classifiers such as RF, AdaBoost, Support Vector Machines (SVM), and Bagging to classify soil nutrient levels by using village-wise fertility indices of Phosphorous Pentoxide (P_2O_5), OC, Fe, and Mn and chemical measurements of EC, pH, Nitrous Oxide (N_2O), Potassium Oxide (K_2O), Sulfate (SO_4, Zn), soil type and recommended fertilizers for crops bajra, cotton, and soybean, based on N_2O, P_2O_5 and K_2O. BRT and RF classifiers were used by Wang et al. [23] to estimate variation in soil OC stock. Khanal et al. [11] adopted site-specific data with *cation exchange capacity, pH, K, Magnesium (Mg), organic matter*, and yield data for corn collected from several fields and developed models using Linear regression, RF, Multilayer Perceptrons, SVM, Gradient Boosting, and Cubist. Ransom et al. [17] incorporated the site-specific soil data to improve Nitrogen fertilizer recommendation for corn and evaluated the performance of Decision Tree, RF classifier, Elastic Net Regression, Stepwise Regression, Ridge Regression, Least Absolute Shrinkage and Selection Operator Regression, Principal Component Regression, and Partial Least Squares Regression. Zhang et al. [24] used an amalgamation of Mutual Information and Ant Colony Optimization to identify total N content spectrometer-sensitive wavebands, and Partial Least Square, Multi Linear Regression, and SVM models were used to determine total nitrogen content. Fernandes et al. [8] verified that Artificial Neural Networks could measure soil organic matter with the highest Accuracy using *pH, Calcium, Mg*, and soil acidity. Zia et al. [25] developed models using M5 tree, Multi Linear Regression, Multilayer Perceptrons, and REPTree to predict total Nitrogen loss. Delavar et al. [6] developed a hybrid model using Artificial Neural Networks and a Genetic Algorithm to predict variation in soil salinity. Mahmoudzadeh et al. [12] estimated OC, using Cubist, K Nearest Neighbours, Extreme Gradient XGBoost, RF, and SVM. Abera et al. [2] predicted variation in N, P, K, and S using RF classifier and recommended fertilizer for wheat based on yield response to fertilizer.

In previous works, soil fertility has been estimated based on a few soil parameters. A few researchers have recommended fertilizer based on a deficit of a few soil nutrients. This research recommends fertilizers based on a deficit of 11 soil chemical parameters to improve soil fertility.

3   Dataset Used

To conduct the experimental study, soil health data [21] collected from the farm-lands of villages of Dakshina Kannada and Gulbarga districts of Karnataka (State), India were used. Dakshina Kannada is a coastal district consisting of acidic soil with pH<6.5. Gulbarga is a noncoastal district consisting of neutral with 6.5<=pH<8.5 or alkaline soil with pH>8.5 [3]. The data collected consists of 19 attributes: sample number, state name, district name, block name, village code, village name, latitude, longitude, and 11 soil chemical parameters. The 11 chemical parameters, namely, EC, pH, OC, P, K, S. B, Cu, Fe, Mn, and Zn, are selected from the dataset. A data cleaning process is performed using the open-source WEKA tool to remove redundant data, missing parameter values, and parameters with a value of 0. The dataset instances are labeled as LOW, MED, or HIGH using the level of soil parameters [22]. After performing data prepro-cessing, the data collected from Dakshina Kannada district (dataset-1) has 36979 instances, with 36796 instances of LOW fertile soil, 158 instances of MED, and 25 instances of HIGH fertile soil. The preprocessing of data collected from the Gulbarga district resulted in 40929 instances, among which 36979 instances were randomly selected (dataset-2). Dataset-2 consists of 36775 instances of LOW fertile soil, 174 instances of MED, and 30 instances of HIGH fertile soil.

4   Proposed Soil Fertility Classifier

The steps involved in the proposed soil fertility classification approach are depicted in Fig. 1. The soil health data is preprocessed and fed as input to the ensemble filter-based feature selection module. The feature scores are calculated for each feature using three different filter-based feature selection techniques, such as InfoG, GainR, and ReliefF. The features are ranked in descending order

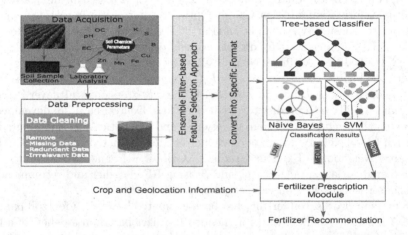

Fig. 1. Sequence of Steps in Proposed Soil Fertility Classification Approach

of the scores obtained, i.e., the feature with the highest score is ranked as 1, whereas the feature with the lowest score is ranked with a higher rank. The feature with rank 1 is considered more relevant, whereas the higher rank is considered least relevant. It was observed that the feature ranks depend on the datasets used. The ranks obtained for a dataset using three different techniques were not constant. The ranking of features based on scores obtained using InfoG, GainR, and ReliefF feature selection approaches are given in Tables 1 and 2, respectively. For dataset-1, it was observed that InfoG and GainR recommended the feature 'S' as least relevant, whereas the least relevant feature recommended by ReliefF was 'B'. For dataset-2, it was found that the least relevant feature recommended by InfoG and GainR was 'P', whereas the least relevant feature recommended by ReliefF was 'S'. To maintain the stability in feature selection, an ensemble filter-based feature selection is proposed as depicted in Fig. 2. The average of ranks obtained by InfoG, GainR, and ReliefF for each soil parameter 'x' is calculated using Eq. (1). The features are arranged based on the ascending order of their rankings. The subset of features by removing the feature with the highest average ranking was given as input to the classifier. The classifier classifies the input data instances as LOW, MED, or HIGH fertility. The soil parameter with the highest average rank was chosen as the parameter to be eliminated.

Table 1. Feature selection scores and ranks obtained for dataset-1

Features (x)	$Score_{InfoG}(x)$	$Score_{GainR}(x)$	$Score_{ReliefF}(x)$	$Rank_{InfoG}(x)$	$Rank_{GainR}(x)$	$Rank_{ReliefF}(x)$	Average Rank
EC	0.00238	0.00193	0.018103	9	8	4	7
pH	0.01207	0.01483	0.055469	2	2	2	2
OC	0.0021	0.00163	0.004477	10	9	7	8.67
K	0.00241	0.00157	0.058884	8	10	1	6.33
P	0.00414	0.00209	0.000824	6	7	9	7.33
S	0.00174	0.00139	0.016596	11	11	5	9
B	0.0071	0.00499	0.000107	4	4	11	6.33
Cu	0.01497	0.02021	0.003144	1	1	8	3.33
Fe	0.00351	0.00311	0.023297	7	6	3	5.33
Mn	0.00641	0.00481	0.012952	5	5	6	5.33
Zn	0.01097	0.01138	0.000107	3	3	10	5.33

Table 2. Feature selection scores and ranks obtained for dataset-2

Features (x)	$Score_{InfoG}(x)$	$Score_{GainR}(x)$	$Score_{ReliefF}(x)$	$Rank_{InfoG}(x)$	$Rank_{GainR}(x)$	$Rank_{ReliefF}(x)$	Average Rank
EC	0.02031	0.27375	0.012372	2	1	2	1.67
pH	0.01665	0.01017	0.032111	4	7	1	4
OC	0.01643	0.00885	0.005453	5	8	5	6
K	0.01397	0.01033	0.00374	8	6	7	7
P	0.00435	0.00297	0.001717	11	11	8	10
S	0.01042	0.00629	0.000475	10	10	11	10.33
B	0.01187	0.00751	0.001393	9	9	9	9
Cu	0.02379	0.04659	0.007797	1	2	3	2
Fe	0.01563	0.01639	0.005312	6	4	6	5.33
Mn	0.01535	0.0242	0.006917	7	3	4	4.67
Zn	0.01828	0.01518	0.001236	3	5	10	6

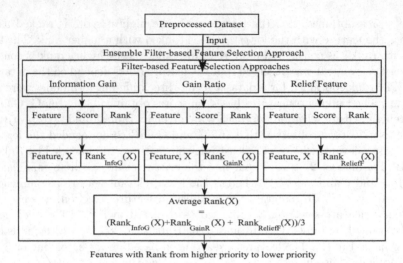

Fig. 2. Proposed Ensemble filter-based feature selection approach

$$Average \quad Rank(x) = \frac{Rank_{InfoG}(x) + Rank_{GainR}(x) + Rank_{ReliefF}(x)}{3} \tag{1}$$

After classifying the instances, a fertilizer suggestion is made based on the value of P and K to boost the amount of 'P' and 'K' in the soil, respectively. The fertility level of N and S is determined based on pH value. The fertility level of N in the soil is considered inadequate when $pH < 5.1$ or $pH > 8.75$. If the pH is between 5.1 and 5.9 or 8 and 8.5, it is considered moderate; otherwise, the nitrogen level is considered high. The fertility level of S is low when $pH < 5.5$, medium when the pH value is between 5.5 and 5.9, and high when $pH >= 5.9$. The deficiency of soil micronutrients, including B, Cu, Fe, Mn, and Zn, is determined based on their chemical measurements, and the fertilizer prescription is made accordingly. Based on the classification results, the proposed model prescribes fertilizer for the specified crops in the given location. This research uses the most prevalent crops in Karnataka state: rice, green gram/ black gram, black pepper, and cucumber. Coastal saline or alluvial soil is used to cultivate the cereal crop paddy. It is grown during the Kharif season (Sowing: June - July, Harvesting: September - October) and the Rabi season (Sowing: October-November, Harvesting: March - April) [18]. Pulses like green gram and black gram are typically cultivated during the Kharif or Zaid (or summer) season in mixed red and black soil. Black pepper is a spice and medicinal crop grown in the Kharif season. The cucumber is widely grown in Zaid/Summer season. Soils often become deficient in major nutrients, particularly N, P, K, and sometimes S and other micronutrients. The quantity of fertilizers to be applied for various crops based on the fertility level of N, P, and K are given in Tables 3, 4, and 5, respectively. Table 6 shows the fertilizer prescription for any crop based on deficiency of 'S' and other micronutrients.

Table 3. Quantity of Neam Coated Urea recommended (kg/ha) based on fertility level of 'N'

	Paddy		Black Pepper	Cucumber	Green gram/ Black gram	
Fertility Level of N	Kharif	Rabi	Kharif	Zaid/ Summer	Kharif	Zaid/ Summer
LOW	289.13	360.87	289.13	173.91	36.96	71.74
MED	217.39	271.74	217.39	130.43	28.26	54.35
HIGH	145.65	180.43	145.65	86.96	17.39	36.96

Table 4. Quantity of Single Superphosphate recommended (kg/ha) based on fertility level of 'P'

	Paddy		Black Pepper	Cucumber	Green gram/ Black gram	
Fertility Level of P	Kharif	Rabi	Kharif	Zaid/ Summer	Kharif	Zaid/ Summer
LOW	418.75	518.75	331.25	418.75	206.25	418.75
MED	312.5	393.75	250	312.50	156.25	312.50
HIGH	206.25	262.5	168.75	206.25	106.25	206.25

Table 5. Quantity of Potassium Chloride recommended (kg/ha) based on fertility level of 'K'

	Paddy		Black Pepper	Cucumber	Green gram/ Black gram	
Fertility Level of K	Kharif	Rabi	Kharif	Zaid/ Summer	Kharif	Zaid/ Summer
LOW	111.67	138.33	310	176.67	55	55
MED	83.33	105	233.33	133.33	41.67	41.67
HIGH	55	70	155	88.33	28.33	28.33

Table 6. Quantity of fertilizers recommended based on deficiency of soil macronutrient 'S' and micronutrients

Soil Parameter	Fertilizer Name	Quantity
S	Sulphur / Gypsum	S: 20-40 kg/ha/ Gypsum:140-280 kg/ha
B	Borax	5-10 kg/ha
Cu	Copper sulphate	5-10 kg/ha
Fe	Ferrous sulphate	25-50 kg/ha
Mn	Manganese sulphate	10 -25 kg/ha
Zn	Zinc sulphate	15 -25 kg/ha

5   Experimental Setup and Results

5.1   Experimental Results of Feature Selection

The feature ranks were obtained using WEKA open-source tool. The proposed approach assigns rank 1 to the feature with the lowest average rank and the highest rank to the feature with the highest average rank. The feature rank

Table 7. Ranking of features using the proposed approach

Dataset-1			Dataset-2		
Features	Average Rank	Rank	Features	Average Rank	Rank
pH	2	1	EC	1.67	1
Cu	3.33	2	Cu	2	2
Fe	5.33	3	pH	4	3
Mn	5.33	4	Mn	4.67	4
Zn	5.33	5	Fe	5.33	5
K	6.33	6	OC	6	6
B	6.33	7	Zn	6	7
EC	7	8	K	7	8
P	7.33	9	B	9	9
OC	8.67	10	P	10	10
S	9	11	S	10.33	11

generated by the proposed approach is given in Table 7. For both datasets, the feature 'S' obtained the highest rank of 11.

5.2   Performance Evaluation of Classifiers

The classifiers and fertilizer prescription module are implemented using Google Collab with python-weka-wrapper3 [16] and java bridge. Two different sets of experiments were conducted 1) by using a 10-fold cross-validation of datasets and 2) by using split datasets, i.e., datasets split into 75%: 25% as training datasets and test datasets, respectively.

The ground truth is established by comparing the actual class to which the instance belongs with the predicted class. The performance of the classifiers is assessed using performance metrics, including Accuracy, Precision, Recall, F1-Score as given in Eq. (2) [1], Eq. (9), Eq. (4), and Eq. (5) [9], respectively.

$$Accuracy = \frac{TP + TN}{TP + TN + FP + FN} \tag{2}$$

where TP, TN, FP, FN indicate True Positive, True Negative, False Positive, and False Negative, respectively. TP indicates an instance for which both predicted, and actual classes are positive. TN is an instance of both predicted and actual classes being negative. FP is an instance for which the predicted class is positive, whereas the actual class is negative. FN refers to an instance where the predicted class is negative, but the actual class is positive.

$$Precision = \frac{TP}{TP + FP} \tag{3}$$

$$Recall = \frac{TP}{TP + FN} \tag{4}$$

$$F1 - Score = \frac{2 * Precision * Recall}{Precision + Recall} \tag{5}$$

The dataset is imbalanced, hence Per-class Precision, Per-class Recall, and Per-class F1-Score using Eq. (6), Eq. (7), Eq. (8) for class i, where 'i' represents LOW, MED or HIGH are calculated.

$$Per-class\ Precision = \frac{TP_i}{TP_i + FP_i} \qquad (6)$$

where i indicates the class LOW, MED, or HIGH.

$$Per-class\ Recall = \frac{TP_i}{TP_i + FN_i} \qquad (7)$$

$$Per-class\ F1-Score = \frac{2*Per-class\ Precision_i*Per-class\ Recall_i}{Per-class\ Precision_i + Per-class\ Recall_i} \qquad (8)$$

The Kappa statistics for each classifier are calculated using Eq. (5.2) [7].

$$Kappa\ statistics = \frac{p_{obs} - p_{exp}}{N - p_{exp}} \qquad (9)$$

where p_{obs} is observed prediction, p_{exp} is the expected prediction and N is the total number of observations.

A batch size of 100 was used to create the tree classifiers. All 11 soil parameters from the datasets were initially considered for the experiment. CART created a tree of size 25 with 13 leaf nodes for dataset-1 and a tree of size 39 with 20 leaves for dataset-2 using number folds for pruning equal to 5 and a seed value of 1. Extra Tree used a seed value of 1, a minimum of two instances at each node for splitting, and randomly selected $\sqrt{(m-1)}$ attributes at each node. For dataset-1, it produced a tree with 1301 nodes, while for dataset-2, it had 561 nodes. Using a fold of three and a seed value of one for dataset-1, J48 produced a tree of 33 and 17 leaves and a tree with a size of 45 with 23 leaves for dataset-2. For both datasets-1 and dataset-2, the RF classifier produced random forests using 100 bags and 100 iterations. For all nodes, REP utilized a minimum variance proportion of 0.001, a maximum number of 3 folds, and a seed value of 1. For dataset-1 and dataset-2, REP produced trees of sizes 27 and 39, respectively. After eliminating the feature 'S' from the datasets, experiments were repeated. CART created a tree of size 25 with 13 leaf nodes for dataset-1 and a tree of size 43 with 22 leaves for dataset-2 using number folds for pruning equal to 5 and a seed value of 1. For datasets-1 and dataset-2, Extra Tree produced trees with 1033 and 655 nodes, respectively. J48 produced a tree with a size of 33 and 17 leaves for dataset-1, creating a tree of size 41 with 21 leaves for dataset-2. The random forest was created using 100 bags and 100 iterations using the RF classifier. For dataset-1 and dataset-2, REP produced trees with sizes of 27 and 45, respectively. NB and SVM classifiers used a batch size of 100. SVM classifier is created using radial bias kernel function, with a seed and cost value of 1, gamma value of 0.1, epsilon of 0.001, loss value of 0.1, and degree 3.

The performance of the classifiers with 10-fold cross-validation for dataset-1 and dataset-2 without removing feature 'S' are given in Tables 8 and 9, respectively. Without eliminating feature 'S', for dataset-1, the Accuracy achieved by

Table 8. Performance of classifiers with 10-fold cross-validation for dataset-1 with all 11 features

Classifier	Accuracy	Per-class Precision			Per-class Recall			Per-class F1-Score			Precision	Recall	F1-Score
		LOW	MED	HIGH	LOW	MED	HIGH	LOW	MED	HIGH			
CART	99.92%	1.000	0.907	0.862	1.000	0.924	1.000	1.000	0.915	0.926	0.999	0.999	0.999
Extra Tree	99.19%	0.996	0.172	0.000	0.996	0.171	0.000	0.996	0.171	0.000	0.992	0.992	0.992
J48	99.92%	1.000	0.900	0.889	1.000	0.911	0.960	1.000	0.906	0.923	0.999	0.999	0.999
RF	99.92%	1.000	0.928	0.885	1.000	0.899	0.920	1.000	0.913	0.902	0.999	0.999	0.999
REP	99.91%	1.000	0.897	0.857	1.000	0.937	0.960	1.000	0.916	0.906	0.999	0.999	0.999
NB	76.99%	1.000	0.017	0.065	0.770	0.911	0.200	0.870	0.033	0.098	0.995	0.770	0.866
SVM	99.51%	0.995	-	-	1.000	0.000	0.000	0.998	-	-	-	0.995	-

Table 9. Performance of classifiers with 10-fold cross-validation for dataset-2 using all 11 features

Classifier	Accuracy	Per-class Precision			Per-class Recall			Per-class F1-Score			Precision	Recall	F1-Score
		LOW	MED	HIGH	LOW	MED	HIGH	LOW	MED	HIGH			
CART	99.84%	0.999	0.886	0.692	1.000	0.805	0.600	0.999	0.843	0.643	0.998	0.998	0.998
Extra Tree	99.64%	0.999	0.661	0.303	0.998	0.695	0.333	0.999	0.678	0.317	0.997	0.996	0.996
J48	99.89%	0.999	0.920	0.742	1.000	0.856	0.767	1.000	0.887	0.754	0.999	0.999	0.999
RF	99.89%	0.999	0.910	0.923	1.000	0.874	0.400	1.000	0.891	0.558	0.999	0.999	0.999
REP	99.89%	1.000	0.895	0.727	1.000	0.879	0.533	1.000	0.887	0.615	0.999	0.999	0.999
NB	88.14%	1.000	0.036	0.018	0.882	0.879	0.167	0.937	0.069	0.032	0.994	0.881	0.932
SVM	99.45%	0.994	-	-	1.000	0.000	0.000	0.997	-	-	-	0.994	-

CART, J48, and RF classifiers was 99.92% whereas REP, Extra Tree, NB, and SVM classifier achieved an Accuracy of 99.91%, 99.19%, 76.99%, and 99.51%, respectively. For dataset-2, the Accuracy attained by J48, RF REP was 99.89% whereas CART, Extra Tree, NB, and SVM classifier was 99.84%, 99.64%, 88.14%, and 99.45%, respectively.

The performance of the classifiers for dataset-1 and dataset-2 after removing feature 'S' are given in Tables 10 and 11, respectively. After removing the feature 'S', for dataset-1 CART, Extra Tree, J48, RF, REP, NB, and SVM classifier achieved an Accuracy 99.92%, 99.20%, 99.92%, 99.93%, and 99.92%, 76.45%, and 99.51%, respectively. For dataset-2, CART, Extra Tree, J48, RF, REP, NB, and SVM classifier achieved an Accuracy of 99.84%, 99.67%, 99.88%, 99.91%, 99.88%, 89.73%, and 99.45%, respectively. The Kappa statistic achieved by classifiers for dataset-1 and dataset-2 is shown in Fig. 3.

The performance of the classifiers with a split dataset (75% as a training dataset and 25% as a test dataset) for dataset-1 and dataset-2 without removing feature 'S' are given in Table 12 and Table 13, respectively. Without removing feature 'S', for dataset-1 CART, Extra Tree, J48, RF, REP, NB, and SVM classifier achieved an Accuracy 99.96%, 99.29%, 99.96%, 99.94%, and 99.95%, 77.06%, and 99.56% respectively. For dataset-2, the CART, Extra Tree, J48, RF, REP, NB, and SVM classifier achieved an Accuracy of 99.81%, 99.62%, 99.87%, 99.88%, 99.83%, 99.84%, and 99.47%, respectively.

Table 10. Performance of classifiers with 10-fold cross-validation for dataset-1 after removing feature 'S'

Classifier	Accuracy	Per-class Precision			Per-class Recall			Per-class F1-Score			Precision	Recall	F1-Score
		LOW	MED	HIGH	LOW	MED	HIGH	LOW	MED	HIGH			
CART	99.92%	1.000	0.907	0.862	1.000	0.924	1.000	0.915	1.000	0.926	0.999	0.999	0.999
Extra Tree	99.20%	0.996	0.195	0.040	0.996	0.196	0.040	0.996	0.196	0.040	0.992	0.992	0.992
J48	99.92%	1.000	0.912	0.862	1.000	0.918	1.000	1.000	0.915	0.926	0.999	0.999	0.999
RF	99.93%	1.000	0.918	0.875	1.000	0.924	0.840	1.000	0.921	0.857	0.999	0.999	0.999
REP	99.92%	1.000	0.897	0.857	1.000	0.937	0.960	1.000	0.916	0.906	0.999	0.999	0.999
NB	76.45%	1.000	0.017	0.068	0.764	0.918	0.200	0.866	0.032	0.101	0.995	0.765	0.862
SVM	99.51%	0.995	-	-	1.000	0.000	0.000	0.998	-	-	-	0.995	-

Table 11. Performance of classifiers with 10-fold cross-validation for dataset-2 after removing feature 'S'

Classifier	Accuracy	Per-class Precision			Per-class Recall			Per-class F1-Score			Precision	Recall	F1-Score
		LOW	MED	HIGH	LOW	MED	HIGH	LOW	MED	HIGH			
CART	99.84%	0.999	0.870	0.750	1.000	0.805	0.600	0.999	0.836	0.667	0.998	0.998	0.998
Extra Tree	99.67%	0.999	0.707	0.231	0.999	0.707	0.200	0.999	0.707	0.214	0.997	0.997	0.997
J48	99.88%	0.999	0.912	0.742	1.000	0.833	0.767	1.000	0.871	0.754	0.999	0.999	0.999
RF	99.91%	0.999	0.933	0.944	1.000	0.885	0.567	1.000	0.909	0.708	0.999	0.999	0.999
REP	99.88%	1.000	0.889	0.679	1.000	0.874	0.633	1.000	0.881	0.655	0.999	0.999	0.999
NB	89.73%	1.000	0.041	0.017	0.898	0.879	0.133	0.946	0.079	0.030	0.994	0.897	0.941
SVM	99.45%	0.994	-	-	1.000	0.000	0.000	0.997	-	-	-	0.994	-

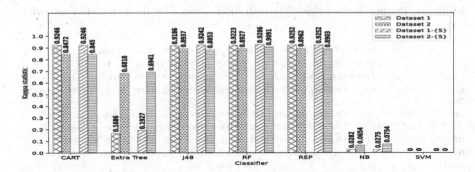

Fig. 3. Kappa statistic achieved by the proposed approach with 10-fold cross-validation

The performance of classifiers after removing feature 'S' for dataset-1 and dataset-2 are given in Table 14 and Table 15, respectively. After removing the feature 'S' for dataset-1 CART, Extra Tree, J48, RF, REP, NB, and SVM classifier achieved an Accuracy 99.96%, 99.32%, 99.96%, 99.96%, 99.95%, 76.27%, and 99.56%, respectively. For dataset-2, CART, Extra Tree, J48, RF, REP, NB, and SVM classifier achieved an Accuracy of 99.85%, 99.73%, 99.88%, 99.90%, 99.81%, 90.14%, and 99.47%, respectively. The Kappa static achieved for dataset-1 and dataset-2 is shown in Fig. 4.

Table 12. Performance of classifiers with Split dataset for dataset-1 with all 11 features

Classifier	Accuracy	Per-class Precision			Per-class Recall			Per-class F1-Score			Precision	Recall	F1-Score
		LOW	MED	HIGH	LOW	MED	HIGH	LOW	MED	HIGH			
CART	99.96%	1.000	0.927	0.667	1.000	0.974	1.000	1.000	0.950	0.800	1.000	1.000	1.000
Extra Tree	99.29%	0.997	0.256	0.000	0.996	0.256	0.000	0.996	0.256	0.000	0.993	0.993	0.993
J48	99.96%	1.000	0.927	0.667	1.000	0.974	1.000	1.000	0.950	0.800	1.000	1.000	1.000
RF	99.94%	1.000	0.946	0.500	1.000	0.897	0.500	1.000	0.921	0.500	0.999	0.999	0.999
REP	99.95%	1.000	0.905	0.667	1.000	0.974	1.000	1.000	0.938	0.800	1.000	0.999	0.999
NB	77.06%	1.000	0.017	0.000	0.770	0.949	0.000	0.870	0.034	0.000	0.996	0.771	0.866
SVM	99.56%	0.996	-	-	1.000	0.000	0.000	0.998	-	-	-	0.996	-

Table 13. Performance of classifiers with Split dataset for dataset-2 with all 11 features

Classifier	Accuracy	Per-class Precision			Per-class Recall			Per-class F1-Score			Precision	Recall	F1-Score
		LOW	MED	HIGH	LOW	MED	HIGH	LOW	MED	HIGH			
CART	99.81%	0.999	0.931	0.500	1.000	0.675	0.778	0.999	0.783	0.609	0.998	0.998	0.998
Extra Tree	99.62%	0.998	0.700	0.154	0.999	0.525	0.222	0.999	0.525	0.182	0.996	0.996	0.996
J48	99.87%	1.000	0.917	0.600	0.999	0.825	1.000	1.000	0.868	0.750	0.999	0.999	0.999
RF	99.88%	0.999	0.897	1.000	1.000	0.875	0.333	1.000	0.886	0.500	0.999	0.999	0.999
REP	99.83%	0.998	0.935	1.000	1.000	0.725	0.667	0.999	0.817	0.800	0.998	0.998	0.998
NB	99.84%	1.000	0.043	0.051	0.920	0.800	0.222	0.958	0.082	0.083	0.994	0.918	0.953
SVM	99.47%	0.995	-	-	1.000	0.000	0.000	0.997	-	-	-	0.995	-

Table 14. Performance of classifiers with Split dataset for dataset-1 after removing feature 'S'

Classifier	Accuracy	Per-class Precision			Per-class Recall			Per-class F1-Score			Precision	Recall	F1-Score
		LOW	MED	HIGH	LOW	MED	HIGH	LOW	MED	HIGH			
CART	99.96%	1.000	0.927	0.667	1.000	0.974	1.000	1.000	0.950	0.800	1.000	1.000	1.000
Extra Tree	99.32%	0.997	0.282	0.000	0.996	0.282	0.000	0.997	0.282	0.000	0.994	0.993	0.993
J48	99.96%	1.000	0.927	0.667	1.000	0.974	1.000	1.000	0.950	0.800	1.000	1.000	1.000
RF	99.96%	1.000	0.927	0.667	1.000	0.974	1.000	1.000	0.950	0.800	1.000	1.000	1.000
REP	99.95%	1.000	0.905	0.667	1.000	0.974	1.000	1.000	0.938	0.800	1.000	0.999	0.999
NB	76.27%	1.000	0.017	0.000	0.762	0.949	0.000	0.865	0.033	0.000	0.996	0.763	0.861
SVM	99.56%	0.996	-	-	1.000	0.000	0.000	0.998	-	-	-	0.996	-

Table 15. Performance of classifiers with Split dataset for dataset-2 after removing feature 'S'

Classifier	Accuracy	Per-class Precision			Per-class Recall			Per-class F1-Score			Precision	Recall	F1-Score
		LOW	MED	HIGH	LOW	MED	HIGH	LOW	MED	HIGH			
CART	99.85%	0.999	0.889	1.000	1.000	0.800	0.778	0.999	0.842	0.875	0.998	0.998	0.998
Extra Tree	99.73%	0.999	0.743	0.375	0.999	0.650	0.333	0.999	0.693	0.353	0.997	0.997	0.997
J48	99.88%	0.999	0.971	0.667	1.000	0.825	0.889	1.000	0.892	0.762	0.999	0.999	0.999
RF	99.90%	0.999	0.923	1.000	1.000	0.900	0.444	1.000	0.911	0.615	0.999	0.999	0.999
REP	99.81%	0.999	0.872	1.000	1.000	0.850	0.667	1.000	0.861	0.800	0.999	0.999	0.999
NB	90.14%	1.000	0.035	0.054	0.902	0.800	0.222	0.949	0.068	0.087	0.994	0.901	0.944
SVM	99.47%	0.995	-	-	1.000	0.000	0.000	0.997	-	-	-	0.995	-

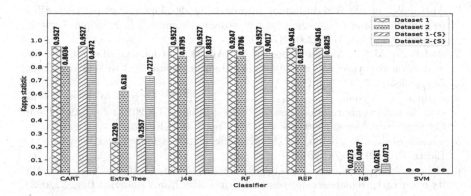

Fig. 4. Kappa statistic achieved by the proposed approach with Split dataset

6   Conclusions and Future Work

Precise classification of soil fertility is a major requirement to enhance agricultural production sustainably. Using a limited number of soil parameters reduces the complexity of classifiers and laboratory chemical analysis costs. An ensemble filter-based feature selection was proposed using three different feature selection approaches: InfoG, GainR, and ReliefF. The proposed approach removes the least relevant feature. The proposed soil fertility classification approach's performance is evaluated using two datasets. The performance of tree-based machine learning classifiers such as CART, Extra Tree, J48, RF, and REP are compared with NB and SVM classifiers. With the elimination of soil parameter 'S' from both the datasets, and using a subset of features consisting of ten soil parameters, *EC, pH, OC, K, P, B, Cu, Fe, Mn, and Zn* the RF classifier outperformed the other classifiers. The RF achieved the highest Accuracy and kappa statistics of 99.96%, 0.9286 for dataset-1 and the highest Accuracy and kappa statistics of 99.90%, 0.9091 for dataset-2, respectively. A significant improvement in kappa statistics were observed after removing the least relevant feature *'S'*, with both 10-fold cross-validation and split dataset. Using both datasets, the RF classifier's performance increased compared to other classifiers after removing feature

'S'. An adequate amount of fertilizers is recommended based on the obtained classification results. In future work, dataset-balancing techniques can be used to obtain accurate results. Real-time soil parameters can be used to assess the proposed approach for classifying soil fertility.

References

1. Abdellatif, A., Abdellatef, H., Kanesan, J., Chow, C.O., Chuah, J.H., Gheni, H.M.: An effective heart disease detection and severity level classification model using machine learning and hyperparameter optimization methods. IEEE Access. **10**, 79974–79985 (2022)
2. Abera, W., et al.: A data-mining approach for developing site-specific fertilizer response functions across the wheat-growing environments in Ethiopia. Experim. Agricult. **58**, S0014479722000047 (2022)
3. Badrinath, M.S., Chidanandappa, H.M., Ali, H.M., Chamegowda, T.C.: Impact of lime on rice yield and available potassium in coastal acid soils of Karnataka. Agropedology **5**, 43–46 (1995)
4. Beraha, M., Metelli, A.M., Papini, M., Tirinzoni, A., Restelli, M. Feature selection via mutual information: new theoretical insights. In: 2019 International Joint Conference on Neural Networks (IJCNN), pp. 1–9. IEEE (2019)
5. Chandrashekar, G., Sahin, F.: A survey on feature selection methods. Comput. Electr. Eng. **40**, 16–28 (2014)
6. Delavar, M.A., Naderi, A., Ghorbani, Y., Mehrpouyan, A., Bakhshi, A.: Soil salinity mapping by remote sensing south of Urmia lake, Iran. Geoderma Reg. **22**, 00317 (2020)
7. DeVellis, R.F.: Inter-rater reliability. In: Encyclopedia of Social Measurement, pp. 317–322. Elsevier (2005)
8. Fernandes, M.M.H., Coelho, A.P., Fernandes, C., da Silva, M.F., Marta, C.C.D.: Estimation of SOM content by modeling with artificial neural networks. Geoderma **350**, 46–51 (2019)
9. Jamali, A., Mahdianpari, M., Brisco, B., Mao, D., Salehi, B., Mohammadimanesh, F.: 3DUNetGSFormer: a deep learning pipeline for complex wetland mapping using generative adversarial networks and Swin transformer. Eco. Inform. **72**, 101904 (2022)
10. Jamali, I., Mohammad B., Shahram J.: Feature Selection in Imbalance data sets. Int. J. Comput. Sci. Issues **9**(3), 42 (2012)
11. Khanal, S., Fulton, J., Klopfenstein, A., Douridas, N., Shearer, S.: Integration of high resolution remotely sensed data and machine learning techniques for spatial prediction of soil properties and corn yield. Comput. Electron. Agric. **153**, 213–225 (2018)
12. Mahmoudzadeh, H., Matinfar, H.R., Taghizadeh-Mehrjardi, R., Kerry, R.: Spatial prediction of SOC using machine learning techniques in western Iran. Geoderma Reg. **1**, e00260 (2020)
13. Osman, K.T.: Soils: principles, properties and management. 1st Edn. Springer Science & Business Media (2012). https://doi.org/10.1007/978-94-007-5663-2
14. Pandey, N.: Role of plant nutrients in plant growth and physiology. In: Hasanuzzaman, M., Fujita, M., Oku, H., Nahar, K., Hawrylak-Nowak, B. (eds.) Plant Nutrients and Abiotic Stress Tolerance. Springer, Singapore (2018). https://doi.org/10.1007/978-981-10-9044-8_2

15. Pes, B.: Ensemble feature selection for high-dimensional data: a stability analysis across multiple domains. Neural Comput. Appl. **32**(10), 5951–5973 (2020)
16. Python-weka-wrapper3. https://pypi.org/project/python-weka-wrapper3/. Accessed 10 Oct 2022
17. Ransom, C.J., et al.: Statistical and machine learning methods evaluated for incorporating soil and weather into corn nitrogen recommendations. Comput. Electron. Agric. **164**, 104872 (2019)
18. Reddy, S.: Cropping seasons of India: Kharif, Rabi, and Zaid(Zayid). https://learnnaturalfarming.com/cropping-seasons-of-india-kharif-rabi-and-zaid/. Accessed 10 Sept 2022
19. Schillaci, C., et al.: Spatiotemporal topsoc mapping of a semi-arid mediterranean region: the role of land use, soil texture, topographic indices and the influence of remote sensing data to modelling. Sci. Total Environ. **601**, 821–832 (2017)
20. Sirsat, M., Cernadas, E., Fernandez Delgado, M., Khan, R.: Classification of agricultural soil parameters in India. Comput. Electron. Agric. **135**, 269–279 (2017)
21. Soil health card india, nutrient status-sample wise (for geo coordinates updation). https://soilhealth.dac.gov.in/PublicReports/nutrientstatussamplesurveywise. Accessed 1 July 2021
22. Sujatha, M., Jaidhar, C.D.: Classification of soil fertility using machine learning-based classifier. In: 2021 2nd International Conference on Secure Cyber Computing and Communications (ICSCCC), pp. 138–143 (2021)
23. Wang, B., et al.: Estimating SOC stocks using different modelling techniques in the semi-arid rangelands of Eastern Australia. Ecol. Ind. **88**, 425–438 (2018)
24. Zhang, Y., Li, M., Zheng, L., Qin, Q., Lee, W.S.: Spectral features extraction for estimation of soil total nitrogen content based on modified ant colony optimization algorithm. Geoderma **333**, 23–34 (2019)
25. Zia, H., Harris, N.R., Merrett, G.V., Rivers, M.: A low-complexity machine learning nitrate loss predictive model-towards proactive farm management in a networked catchment. IEEE Access **7**, 26707–26720 (2019)

Privacy-Preserving Pest Detection Using Personalized Federated Learning

Junyong Yoon, Ajit Kumar, Jaewon Jang, Jaeheon Kim,
and Bong Jun Choi(✉)

Soongsil University, Seoul, South Korea
davidchoi@soongsil.ac.kr

Abstract. Today machine learning and deep learning are being used in all industrial and social reform. The use and acceptability of AI solutions are rapidly growing. So, traditional agriculture practices adopt these modern technologies and move towards precision farming. Pest detection and classification is one of the critical areas of concern for agriculture and farmers. Deep learning-based detection and classification have recently automated the process, making detection significantly faster. However, centralized training is required to upload crop images (infected or not infected) which leads to privacy invasion and may lead to a negative reputation for the crop. It may incur financial losses due to the low pricing of the harvest to the farmer. Therefore, we provide privacy-preserving pest detection and classification using personalized federated learning that generates detection models based on agricultural characteristics while the farmers keep the data. We provide a performance comparison between centralized and federated approaches for five classes of pests. Further, we perform experiments to create a group-based personalized model. Through experiments, we found that the accuracy of the federated approach is lower than the centralized training. We also found that the performance of groups varies in personalized FL, so the accuracy of Group A (0.69) is higher than Group B (0.63). Based on the experimental result, the proposed solution is suitable for agriculture because of the privacy preservation and personalization with distributed and low computing devices.

Keywords: Deep learning · Precision Agriculture · Privacy-preserving computing · Federated Learning · Personalized Federated Learning

1 Introduction

With the advancement in Information and Communication Technology (ICT), traditional agriculture is evolving into smart agriculture. Along with ICT, the adoption of artificial intelligence (AI) has provided other opportunities to help and improve the agriculture process by easing soil testing to disease prediction in the crop. Today, there is wide adoption of deep learning and machine learning technologies in agriculture, and farmers can use them due to the affordable Internet of Things (IoT) devices ubiquitously connected to smartphones.

M. K. Saini et al. (Eds.): ICA 2023, CCIS 1866, pp. 58–70, 2023.
https://doi.org/10.1007/978-3-031-43605-5_5

Deep learning is suitable for classification and prediction, so there are many use cases in agriculture [7,10,13]. The crop and fruit production forecast can be made using deep learning [12], and it can also be used for plant classification (for plant phenology) [18] and disease identification in the plant [2,17,21]. Some other applications are like weed detection [20], species recognition [9], and water management [16]. Considering the impact of deep learning on agriculture and humans, there has been rising interest in research and development in agriculture using deep learning [6].

Deep learning is helpful and solves many challenges for traditional agriculture [19]. However, the existing deep learning-based solutions have many bottlenecks and limitations, such as lower prediction or classification accuracy and high computational and memory requirements for training and inference [23]. In addition to performance and hardware-based limitations, there are some key challenges, such as the noisy background in crop images, conditions of image capture (controlled vs. natural environment), complex segmentation (a symptom of disease in plant and fruit), and size of the dataset (difficulty in collecting images [19]) in preparing datasets and in training deep learning models [3].

Restriction for collecting data due to privacy law adds another challenge to the already smaller dataset of developing a deep learning model for smart agriculture [2,19]. The model trained with a smaller dataset lacks generalization and is prone to errors, while inference is made on the validation dataset or with a natural environment.

Traditionally, the deep learning models are trained in a centralized manner, i.e., all the training data from multiple sources. For example, data from multiple farms or controlled environment agriculture (CEA) facilities are collected at a central server that performs pre-processing, labeling, and training. Such an approach is computationally and communication inefficient and rooting to data privacy, leading to data scarcity. Recently, Federated Learning (FL), a distributed training architecture that does not require centralized data collection, has been adopted to address the bottlenecks of traditional centralized deep learning approaches [14]. The main goal of FL is to achieve data privacy for clients or data sources and yet to be able to train models. The key benefit of FL is that raw data never leaves the source/origin, and only a lower dimension representation (i.e., model weight) is shared with the parameter server. In most cases, the parameter server is either among available devices or trusted. However, in the case of external or untrusted parameter servers, methods such as differential privacy and multi-parties computation further improve the privacy and security of model weight [4,24].

As discussed, collecting crop data at a centralized location raises privacy concerns and limits data access. In addition, farmers will be more concerned about sharing information about plant diseases due to fear of unnecessarily hurting their crop costs. Such use cases undermine the usability and training of centralized deep learning models. Considering the challenges of centralized deep learning and the benefits of FL, the proposed work presents the performance of FL for pest classification. Using personalized federated learning (PFL), the proposed work also addresses the requirement of the specialized model that would be required due to differences in types of crops and plants (morphology), diseases (pathology), or changes in the environment (natural vs. controlled or tropical vs. temperate vs. arid vs. alpine, etc.). In PFL, the clients participate in learning similar to vanilla FL (non-PFL); however, they retrain the global model and create a personalized model suitable for their needs. We have trained deep learning models in centralized and federated learning approaches and compared their performance. In the proposed work, we have made the following contributions:

- Proposed a personalized federated learning model to detect pests on the farms.
- Proposed to create clusters based on terrain features and temperature and humidity environments and assign farms with similar features to the same cluster group. Each farmer participates in FL. They create a local model with their dataset and send the local model to the central server. The server aggregates the collected local models to create a global model. The global model created by running FL per group is distributed to each clustered group.
- We conducted an experiment using the pest dataset on Kaggle and compared the accuracy of Centralized Machine Learning (CML) and Federated Learning (FL), and then compared the accuracy of FL when the dataset is IID (general environment) and non-IID (personalized environment). From the experimental results, the accuracy of CML is 72%, the accuracy of general FL is 49%, and the accuracy of FL in the personalized environment is 69%.
- As a result, the personalized FL outperforms the traditional FL under a non-IID environment and achieves accuracy close to that of the CML.

The remaining article is structured as follows. Section 2 presents the related works on using deep learning and federated learning for plant and disease classification. The methodologies of the proposed work are explained in Sect. 3. Section 4 presents experiments and performance evaluation details. We have concluded the proposed work in Sect. 5.

2   Related Works

Deep learning can be used for different related tasks of agriculture. In this related work section, we have considered discussing and analyzing the literature on classification tasks such as types of plant diseases or morphology. We have also included early works of federated learning in agriculture. Too et al. [22] presented a review of fine-tuning and evaluation of convolutional neural network (CNN) deep learning architectures (VGG net, ResNet, and DenseNet) for plant disease classification using the leaf image dataset from plantVillage [8]. The dataset has images of fruit plants' healthy and unhealthy (variation like early_blight, healthy, and late_blight) leaves; each fruit has unequal classes depending on plant diseases. The authors achieved the best performance of 99.75% using DenseNets. Saleem et al. [21] presented a comprehensive survey of deep learning architectures, data augmentation, and features for plant disease detection using the plantVillage dataset. Arsenovic [2] introduced a plant disease dataset recorded in an actual environment (different weathers, angles, daylight, and varing backgrounds) and improved by classical and using generative adversarial networks. Through the review of key literature, authors have mentioned four limitations of existing deep learning-based disease classification: i) data scarcity, ii) effectiveness in actual conditions, iii) accuracy, and iv) identifying the stage of the disease. The authors have highlighted the issue of performance degradation when the model trained with images collected in a controlled environment is tested on the actual environment.

Smart agriculture will significantly benefit from deep learning; Ale et al. [1] proposed a deep learning-based solution for plant disease detection. The authors aim to reduce the size of the trained model and enable it to run on Internet of Things (IoT) devices using the proposed lightweight deep learning network (by reducing neurons in each layer and input size) and transfer learning. Geetharamani et al. [8] conducted experiments using the plantVillage[1] dataset and achieved 71.98% and 90% accuracy with input sizes 32×32 and 512×512, respectively. Mohanty et al. [17] used a deep learning-based smartphone app for disease diagnosis, and the model was trained to classify 14 crop species and 26 diseases.

Recently, Khan et al. [11] used FL and UAV to classify different pests. The authors have trained EfficientNet architecture on the angle-based augmented dataset to classify pests into nine classes. The local model is trained, and updates are collected from four locations. Deng et al. [5] have used R-CNN with ResNet in the FL approach to improve the convergence and training speed. The authors have considered the detection of a number of pests of different sizes and diseases in fruit. The proposed work is similar to Khan et al. [11] and Deng et al. [5] in the use of FL for agriculture; however, our approach is focused on personalization, whereas Khan et al. [11] mainly aim to use UAV-based classification and Deng et al. [5] addressing the optimization issue in terms of training speed and convergence.

We can observe from the related works that earlier deep learning models were being trained only in a centralized approach. Recently, a few FL-based methods have been adopted for agriculture tasks. However, the existing FL-based techniques mainly focus on adopting FL for agricultural tasks and improving the performance of training and inference time performance. Considering the research gap, the proposed work has used PFL to fulfill the need for a personalized model for the agricultural task that can address plant diversity in terms of morphology and pathology. In the further section, the proposed work is presented in detail.

---

[1] https://www.kaggle.com/datasets/emmarex/plantdisease.

3 Pest Classification Using Federated Learning

3.1 Federated Learning

In this study, we use associative learning for pest detection. Federated learning was first published in 2015 by McMahan et al. [15]. It is characterized by two main advantages: improved privacy and communication efficiency. Unlike centralized learning, which requires tens of thousands of local devices to transmit data to a centralized server, resulting in high network traffic and storage costs, federated learning can significantly reduce communication costs by transmitting only the results of learning from local clients (local models) to the server.

The motivation for using federated learning for pest classification is to ensure the privacy-preservation of farm-owned data. Collecting data for pest detection inevitably reveals the pest status of the farm where the data was collected. However, since this data is directly related to the farmer's income, most farmers do not want to reveal this data. Traditional pest detection algorithms collect pest data from each farmer to create a model. This reveals information about pest-rich farmers (i.e., farmers with poor-quality produce). To solve this problem, we propose a pest detection algorithm using association learning to generate a pest detection model without revealing data on pest status.

3.2 Proposed Algorithm

We first provide the overview of our privacy-preserving pest detection algorithm using personalized federated learning in steps, as illustrated in Fig. 1.

Step 1: A number of clusters are created based on features that affect pest reproduction, such as terrain, temperature, and humidity. Then, the server computes the similarity of the features of each farmer (hereinafter referred to as a *client*) to that of the clusters. Then, the server assigns the client to the clusters that share the most similar features.

Step 2: Each client participates in associative learning with their own pest dataset. Each client belonging to the same cluster creates a local model by training from their dataset and sends it to the central server.

Step 3: The central server collects local models from each client. Since the collected models are categorized by the cluster to which each client belongs, the models from the clients in the same cluster are aggregated to create a global model for the cluster.

Step 4: The server distributes the global models created for different clusters to respective clusters. Hence, the model distributed to the same cluster is shared by all clients in the cluster.

Finally, the above steps can be repeated until we obtain a satisfactory model for each cluster. In general, the performance is expected to improve with the iterations as more data is trained.

The details of the proposed algorithm are shown in Algorithm 1. We assume that there are N clients and K clusters, where each client is denoted as $i \in \{1, 2, \ldots, N\}$ and each cluster is denoted as $k \in \{1, 2, \ldots, K\}$.

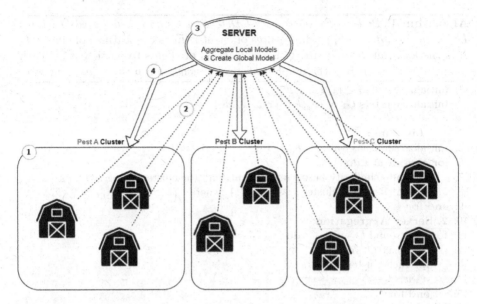

Fig. 1. Proposed Personalized Federated Learning Model for Pest Detection: (1) the server creates a group of farms with similar environments, (2) each farm locally trains the model using its own data and sends the updated model to the server, (3) the server aggregates the models received from the farms to create an updated global model for each cluster, and (4) the server distributes the updated model to each group of farms belonging to the same cluster.

Clustering is presented in lines 5-9. First, the server computes the similarity between the features of the client F_i and the features of cluster F_k, and selects the combination of the client and the cluster that gives the highest similarity defined as $(i, k) = \text{argmax}_{i,k} S(F_i, F_k)$, where $S(F_i, F_k)$ represents the function that computes the similarity between the client i and the cluster k. The similarity can be computed in various ways. Note that the features used in the computation of similarity, such as terrain, temperature, and humidity, are general information that does not contain sensitive private information. So, the client can safely share this information with the server to be assigned to the most appropriate cluster that shares the most similar features.

The server-side execution is presented in lines 11-16. The server sends the initial weight w^0 to the clients. In each iteration, the server collects local model updates from the m participating clients. The updated model received from each client w_i^t is aggregated to create an updated global model for each cluster w_k^{t+1}. The local model received from the clients is weighted proportional to the number of data that each client contributed to training the model. Note that, in line 12, the $max(\cdot)$ function is used to ensure that there is at least one client participating in the process.

Algorithm 1. *Privacy-Preserving Pest Detection Using Personalized Federated Learning Algorithms.* F_i is the features of client i, F_k is the features of cluster k, B is the local mini-batch size, C is the fraction of clients to participate in each round $(0 \leq C \leq 1)$, E is the number of local epochs, and η is the learning rate.

1: Initialize clients $i \in \{1, 2, \ldots, N\}$
2: Initialize clusters $G_k = \{\emptyset\}, k \in \{1, 2, \ldots, K\}$
3: initialize w^0
4: **1. Clustering:**
5: Initialize similarity value $S(F_i, F_k)$
6: **for** each client i **do**
7:     Compute similarity between the clients and the clusters using $S(F_i, F_k)$
8:     Assign client i to cluster k as $G_k \cup \{i\}$, where $(i, k) = \text{argmax}_{i,k} S(F_i, F_k)$
9: **end for**
10: **2. Server Aggregation:**
11: **for** each round $t = 1, 2, \ldots$ **do**
12:     $m \leftarrow \max(C \cdot N, 1)$
13:     **for** each cluster k **do**
14:         $w_k^{t+1} \leftarrow \sum_{i \in G_k} \frac{n_i}{n} w_i^t$
15:     **end for**
16: **end for**
17: **3. Client Update:**
18: $\mathcal{B} \leftarrow$ (split $\mathcal{P}_i$ into batches of size B based on cluster k recommendation)
19: **for** each local epoch j from 1 to E **do**
20:     **for** batch $b \in \mathcal{B}$ **do**
21:         $w_i^t \leftarrow w_i^t - \eta \nabla \ell(w; b)$
22:     **end for**
23: **end for**
24: return w_i^t to server

The client-side execution is presented in lines 18-24. Each client i divides its local data $\mathcal{P}_i$ into mini-batches of size B based on the recommendation of cluster k and calculates the model weights w_i^t by training the model using. Then, the updated gradient is sent to the server.

4 Performance Evaluation

4.1 Experimental Setup

The dataset used in this experiment was processed from the Pest Dataset[2] on Kaggle. The original dataset consists of 9 classes; aphid, armyworm, beetle, bollworm, grasshopper, mites, mosquito, sawfly, and stem borer, with 300 training and 50 testing images per class. We cleaned this dataset by reducing the number of classes from 9 to 5, merging training and test images, and removing duplicated images. As a final step, they were augmented to complement the

---

[2] https://www.kaggle.com/datasets/simranvolunesia/pest-dataset.

insufficient number of images. Moreover, based on our assumption about heterogeneous environments, we generated two non-IID data groups; the number of images of a class is higher than others. Figure 2 is an example of this process.

IID data is "independent" and "identically distribution" data. Non-IID data is "non-independent" and "unequally distributed" data. Referring to the Fig. 3, IID data is a dataset of data of the same size with the same distribution, and Non-IID data is a dataset of different sizes and different distributions.

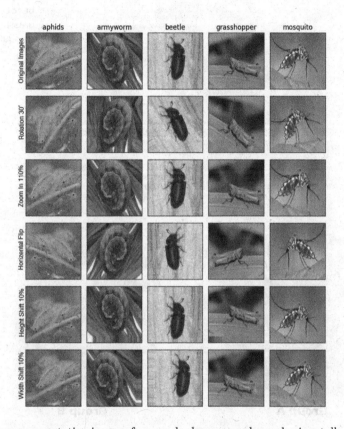

Fig. 2. Five representative images from each classes are shown horizontally, and their augmented version is shown below the respective images. We used the data cleansing process to make the data cleaner overall. The preprocessed version reduces the unnecessary parts of the data imported from Kaggle into 5 classes, and the augmented version is the result of augmenting the quality data by reducing the unnecessary parts.

The experimental environment used Colab, provided by Google. Colab does not require any environment configuration, and anyone can easily use the GPU environment for free. The experimental settings are Centralized ML, FL with general IID, FL with Group A, and FL with Group B. The experiments are conducted by comparing Centralized ML and FL with general IID. Group A and B augment the amount of specific pest data to assume that there are more clusters with specific pest data. We used the CNN base model as the base detection model for CML and FL.

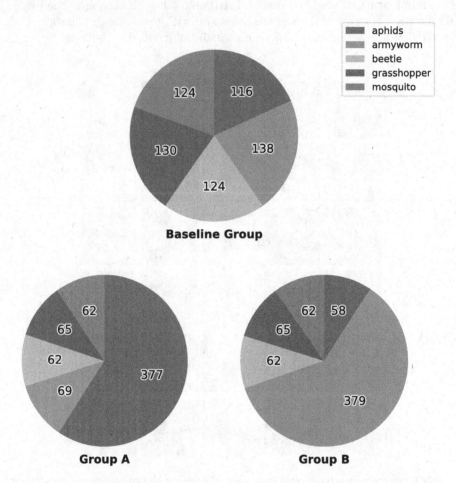

Fig. 3. The percentage of datasets used in the experiment. The Baseline group is a dataset with data for each pest divided evenly (IID data). Group A and B are datasets with increased data for different specific pests, creating datasets specialized for specific pests (Non-IID data).

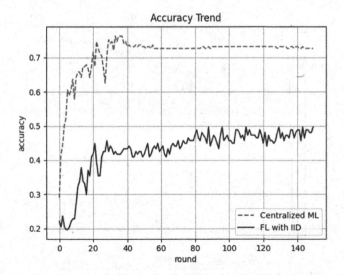

Fig. 4. Comparison of accuracy between Centralized Machine Learning(CML) and Federated Learning(FL). CML has an accuracy of 0.72 and FL has an accuracy of 0.49.

4.2   Results

Figure 4 shows the comparison of accuracy in each round between Centralized Machine Learning and Federated Learning with IID. Here, round means one round of the FL algorithm, i.e., the process until the global model is created. The accuracy obtained using Centralized Machine Learning is 0.72, while the accuracy result of Federated Learning with IID is 0.49. In general, Federated Learning is less accurate than Centralized Machine Learning because it only receives the weight of each client to create a model, whereas Centralized Machine Learning collects all data and learns directly to create a model, so it is inevitably less accurate than Centralized Machine Learning.

Figure 5 shows the comparison of accuracy between the FL with IID data and FL with Group A and B. The accuracy result of FL with IID is 0.49, which is the same as the previous result, and the results of personalized FL with Group A and B are higher at 0.69 and 0.63, respectively. The results show that FL models personalized for specific datasets can achieve better accuracy than the traditional FL models using general datasets.

By analyzing the results compared, we can observe that general Federated Learning is less accurate than Centralized ML, but FL using datasets specialized for specific data can make the accuracy of detecting specific data close to the accuracy of Centralized Machine Learning. In other words, the personalized Federated learning method for each cluster, such as the algorithm introduced earlier, produces higher accuracy than learning with general Federated Learning.

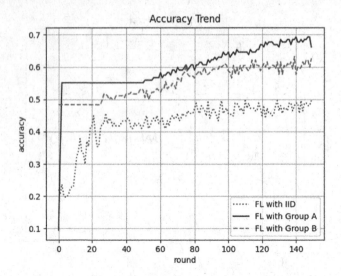

Fig. 5. Comparison of accuracy between IID data and non-IID data. FL with IID data has an accuracy of 0.49, and FL with non-IID has an accuracy of 0.69 and 0.63, respectively.

5   Conclusion

The use of deep learning and machine learning is driving today's smart agriculture. We have proposed the PFL for training and building a classification model for pest classification in personalized and privacy-preserving mode. Thus, the proposed work has addressed three critical issues of traditional deep learning, i.e., i) data scarcity, ii) data privacy, and iii) the need for a personalized model. The experimental results (0.69 and 0.63 accuracies) of the PFL model in the non-IID setting are encouraging and open possibilities for further investigations and improvements. With the proposed work, we have experimented with the capability of FL and PFL in opposition to centralized learning. In continuation of the proposed work, in future work, we aim to improve the performance of PFL further and extend the classification to plant and fruit disease classification. We will address the morphological and pathological challenges that increase sample heterogeneity and create many classes.

Acknowledgement. This research was supported by the MSIT Korea under the NRF Korea (NRF-2022R1A2C4001270) and the Information Technology Research Center (ITRC) support program (IITP-2022-2020-0-01602) supervised by the IITP, and the KIAT grant funded by the Korean government (MOTIE) (P0017123, The Competency Development Program for Industry Specialist).

References

1. Ale, L., Sheta, A., Li, L., Wang, Y., Zhang, N.: Deep learning based plant disease detection for smart agriculture. In: 2019 IEEE Globecom Workshops (GC Wkshps), pp. 1–6. IEEE (2019)
2. Arsenovic, M., Karanovic, M., Sladojevic, S., Anderla, A., Stefanovic, D.: Solving current limitations of deep learning based approaches for plant disease detection. Symmetry **11**(7), 939 (2019)
3. Barbedo, J.G.A.: A review on the main challenges in automatic plant disease identification based on visible range images. Biosys. Eng. **144**, 52–60 (2016)
4. Byrd, D., Polychroniadou, A.: Differentially private secure multi-party computation for federated learning in financial applications. In: Proceedings of the First ACM International Conference on AI in Finance, pp. 1–9 (2020)
5. Deng, F., Mao, W., Zeng, Z., Zeng, H., Wei, B.: Multiple diseases and pests detection based on federated learning and improved faster R-CNN. IEEE Trans. Instrum. Meas. **71**, 1–11 (2022)
6. Dokic, K., Blaskovic, L., Mandusic, D.: From machine learning to deep learning in agriculture-the quantitative review of trends. In: IOP Conf. Ser. Earth Environ. Sci. **614**, 012138 (2020). IOP Publishing (2020)
7. Gandhi, R.: Deep reinforcement learning for agriculture: principles and use cases. In: Reddy, G.P.O., Raval, M.S., Adinarayana, J., Chaudhary, S. (eds.) Data Science in Agriculture and Natural Resource Management. SBD, vol. 96, pp. 75–94. Springer, Singapore (2022). https://doi.org/10.1007/978-981-16-5847-1_4
8. Geetharamani, G., Pandian, A.: Identification of plant leaf diseases using a nine-layer deep convolutional neural network. Comput. Electr. Eng. **76**, 323–338 (2019)
9. Grinblat, G.L., Uzal, L.C., Larese, M.G., Granitto, P.M.: Deep learning for plant identification using vein morphological patterns. Comput. Electron. Agric. **127**, 418–424 (2016)
10. Jagtap, S.T., Phasinam, K., Kassanuk, T., Jha, S.S., Ghosh, T., Thakar, C.M.: Towards application of various machine learning techniques in agriculture. Mater. Today Proceed. **51**, 793–797 (2022)
11. Khan, F.S., et al.: Federated learning-based UAVs for the diagnosis of plant diseases. In: 2022 International Conference on Engineering and Emerging Technologies (ICEET), pp. 1–6. IEEE (2022)
12. Khan, T., Sherazi, H.H.R., Ali, M., Letchmunan, S., Butt, U.M.: Deep learning-based growth prediction system: a use case of china agriculture. Agronomy **11**(8), 1551 (2021)
13. Liakos, K.G., Busato, P., Moshou, D., Pearson, S., Bochtis, D.: Machine learning in agriculture: a review. Sensors **18**(8), 2674 (2018)
14. McMahan, B., Moore, E., Ramage, D., Hampson, S., y Arcas, B.A.: Communication-efficient learning of deep networks from decentralized data. In: Artificial intelligence and statistics, pp. 1273–1282. PMLR (2017)
15. McMahan, H.B., Moore, E., Ramage, D., y Arcas, B.A.: Federated learning of deep networks using model averaging. CoRR abs/1602.05629 (2016), http://arxiv.org/abs/1602.05629
16. Mehdizadeh, S., Behmanesh, J., Khalili, K.: Using MARS, SVM, GEP and empirical equations for estimation of monthly mean reference evapotranspiration. Comput. Electron. Agric. **139**, 103–114 (2017)
17. Mohanty, S.P., Hughes, D.P., Salathé, M.: Using deep learning for image-based plant disease detection. Front. Plant Sci. **7**, 1419 (2016)

18. Nesteruk, S., et al.: Image compression and plants classification using machine learning in controlled-environment agriculture: Antarctic station use case. IEEE Sens. J. **21**(16), 17564–17572 (2021)
19. Ojo, M.O., Zahid, A.: Deep learning in controlled environment agriculture: a review of recent advancements, challenges and prospects. Sensors **22**(20), 7965 (2022)
20. Pantazi, X.E., Tamouridou, A.A., Alexandridis, T., Lagopodi, A.L., Kashefi, J., Moshou, D.: Evaluation of hierarchical self-organising maps for weed mapping using UAS multispectral imagery. Comput. Electron. Agric. **139**, 224–230 (2017)
21. Saleem, M.H., Potgieter, J., Arif, K.M.: Plant disease detection and classification by deep learning. Plants **8**(11), 468 (2019)
22. Too, E.C., Yujian, L., Njuki, S., Yingchun, L.: A comparative study of fine-tuning deep learning models for plant disease identification. Comput. Electron. Agric. **161**, 272–279 (2019)
23. Wani, J.A., Sharma, S., Muzamil, M., Ahmed, S., Sharma, S., Singh, S.: Machine learning and deep learning based computational techniques in automatic agricultural diseases detection: methodologies, applications, and challenges. Arch. Comput. Methods Eng. **29**(1), 641–677 (2022)
24. Wei, K., et al.: Federated learning with differential privacy: algorithms and performance analysis. IEEE Trans. Inf. Forensics Secur. **15**, 3454–3469 (2020)

A Review on Applications of Artificial Intelligence for Identifying Soil Nutrients

Shagun Jain[(⊠)] and Divyashikha Sethia

Department of Software Engineering, Delhi Technological University, Delhi, India
shagunjain191172@gmail.com, divyashikha@dtu.ac.in

Abstract. Soil nutrients play a crucial role in the growth and productivity of crops, which are essential for meeting the increasing global food demand. However, the traditional methods of soil nutrient analysis are time-consuming, labour-intensive, and often not cost-effective. Recently, there has been a growing interest in applying Artificial Intelligence (AI) techniques to soil nutrient analysis. AI can help optimize soil nutrient management and improve crop yields. Farmers can identify potential nutrient deficits in soil quality using artificial intelligence technology, particularly electronic applications for Machine Learning (ML) and Deep Learning (DL). This paper aims to summarise and assess state of the art in using Artificial Intelligence for soil nutrient analysis. The study explores the different ML and DL techniques used for soil nutrient analysis and the results obtained from their application. The review additionally points out areas of improvement for the current research and suggests possibilities for future investigations. It further explains different datasets based on satellite images, smartphone images, and chemical data. Furthermore, the report includes several publicly available soil datasets.

Keywords: Soil nutrients · Artificial Intelligence · Machine Learning · Deep Learning

1 Introduction

Soil is a vital resource for the Earth but still needs to be appreciated. The health of the soil has significantly impacted the state of the food chain, the quality of air and water, and many other ecosystems on Earth. Earth's population is overgrowing, and soil nutrient analysis is essential to feed such a vast population in precision farming.

Soil nutrients heavily influence the yield of crops [3]. The primary source of nutrient absorption for plants is the soil. By understanding the nutrient status of the soil, farmers can take appropriate measures to adjust the levels of deficient nutrients and maintain optimal soil conditions for plant growth. It can help increase crop yield and improve the quality and consistency of the crops. Soil nutrient prediction improves agricultural productivity, environmental sustainability, and food security. Farmers, agronomists, and soil scientists

© The Author(s), under exclusive license to Springer Nature Switzerland AG 2023
M. K. Saini et al. (Eds.): ICA 2023, CCIS 1866, pp. 71–86, 2023.
https://doi.org/10.1007/978-3-031-43605-5_6

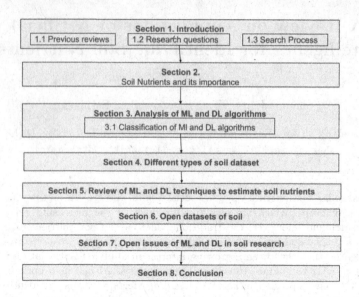

Fig. 1. Structure of Paper

Table 1. Comparison of Review Studies. [N1]-Macronutrients, [N2]-Micronutrients, [N3]-Soil pH, [N4]-Soil Organic Carbon (SOC), [N5]-Soil Organic Matter (SOM)

Study	ML	DL	N1	N2	N3	N4	N5	Open Issues
Odebiri et al. [21]	×	✓	×	×	×	✓	×	✓
Wankhede et al. [42]	✓	×	✓	×	✓	×	×	×
Shahare et al. [29]	✓	×	✓	✓	✓	✓	✓	×
Lamichhane et al. [16]	✓	×	×	×	×	✓	×	×
Proposed Study	✓	✓	✓	✓	✓	✓	✓	✓

widely use it to inform soil management and crop production decisions. Lack of artificial intelligence (AI) usage to anticipate soil nutrient levels may lead to problems such as low crop yields, access restrictions to testing tools, inaccurate testing procedures, and inefficient fertilizer use. The field of AI is expansive and includes a diverse range of technologies and methodologies, such as Machine Learning (ML) and Deep Learning (DL). The past ten years have seen a broad application of ML approaches in various scientific domains [24]. DL applications mainly minimize the dependence on spatial-form designs and preprocessing techniques by simplifying the entire process [32]. Artificial Intelligence's (AI) adaptability, high performance, accuracy, and cost-effectiveness are critical agricultural concepts [7]. Previously inaccessible agricultural data sources, like satellite and Unmanned Aerial Vehicle (UAV) readings, humidity sensor readings, and ground-based weather stations, are now available to agricultural producers for decision-making. Figure 1 depicts the structure of this paper.

1.1   Previous Reviews

This section presents the existing survey papers about the methods of soil nutrient analysis. In their study, Odebiri et al. [21] conducted a concise overview of how Neural Networks (NN) and DL methods can be applied to predict soil organic carbon (SOC) levels using remotely sensed data. Wankhede et al. [42] explain several machine-learning techniques for analyzing soil pH, nitrogen, phosphorus, and potassium. Shahare et al. [29] reviewed various machine-learning techniques to evaluate crop yield and soil properties. Lamichhane et al. [16] presented a review identifying the RF method as the most significant concerning digital soil organic carbon mapping.

It was deemed necessary to compile a survey of research publications describing the uses of artificial intelligence approaches for soil due to advancements in Soil Artificial Intelligence. This review presents the most recent computational studies on soil nutrient analysis. Table 1 compares previous literature studies and this study.

1.2   Research Questions

This study's primary goal was to examine whether ML and DL methods are being used successfully in the field of soil. As a result, this study addresses the following research questions:

- *RQ1:* What are the different types of soil datasets for ML and DL techniques?
- *RQ2:* Which ML models are applied to estimate soil nutrients?
- *RQ3:* Which DL models are applied to estimate soil nutrients?
- *RQ4:* What are the open challenges while predicting soil nutrients using ML and DL techniques?

RQ1 examines various types of datasets of soil used for ML and DL techniques. RQ2 investigates how ML is used in the soil domain and identifies the most popular techniques. RQ3 examines various DL methods for predicting soil nutrient levels, and RQ4 outlines the significant challenges of DL and ML in analyzing soil nutrients.

1.3   Search Process

The literature review on artificial intelligence and soil was diverse and probed at various depths. The database is Google Scholar and Scopus. The following search strings are used to gather data "soil AND nutrient AND analysis AND artificial AND intelligence," "machine learning OR deep learning AND soil pH OR soil organic carbon OR soil organic matter," and "soil AND hyperspectral AND images AND deep learning OR machine learning." The titles and abstracts of each paper were collected. Research articles over the preceding four years (2019-2021) that deal with soil nutrient analysis using ML and DL models are considered for this study.

This paper examines the various model evaluation criteria and explores how well they function with soil-specific ML and DL techniques. The article has furnished the following details in the subsequent sections. An introduction to soil nutrients and their importance is mentioned in Sect. 2. Section 3 shows an analysis of ML and DL algorithms. Section 4 describes different types of soil datasets. The literature review results to respond to the research questions are presented in Sect. 5. Some soil domain open datasets are provided in Sect. 6. Section 7 describes various available issues in ML, and DL approaches for analyzing soil nutrients, and a summary is provided in Sect. 8.

2   Soil Nutrients and Its Importance

Soil nutrients are mainly classified into two categories which are macronutrients and macronutrients. Plants require macronutrients such as Nitrogen(N), Phosphorous(P), Potassium(K), Sulfur(S), Calcium(Ca), and Magnesium(Mg) in relatively high quantities. They also need micronutrients such as Chlorine(Cl), Iron(Fe), Boron(B), Manganese(Mn), Zinc(Zn), Copper(Cu), and Molybdenum(Mo), but in much lesser amounts. The growth of crops relies on several other crucial soil nutrients, which are listed below [5]:-

1. *Soil pH:* The soil's pH determines the soil's acidity or alkalinity. For optimal plant growth, the soil pH should generally fall within 5.5 to 7.5.
2. *Soil Organic Carbon (SOC):* SOC exclusively refers to the carbon atom in organic molecules. It is essential for synthesizing the organic acids in soil, which are necessary for dissolving soil minerals, their availability to plants, and the leaching of nutrients.
3. *Soil Organic Matter (SOM):* SOM is primarily made up of carbon, hydrogen, and oxygen. At the same time, organic wastes also contain trace amounts of nitrogen, phosphorus, sulfur, potassium, calcium, and magnesium.

Although these nutrients are necessary for plants, they must be provided in the right quantities. These nutrients are required for soil fertility to grow, resulting in crop nutrient demands needing to be fulfilled. Nutrient excess or deficiency can harm a plant's growth and development. For crops to develop well and firmly, the soil has to contain a balance of all nutrients.

3   Analysis of ML and DL Algorithms

Both ML and DL algorithms are used to predict soil nutrients. ML applications employ supervised and unsupervised learning techniques to improve data analysis and generate sufficient data to facilitate statistical solutions. DL techniques help to process a large amount of data to test soil nutrients. It is essential to do it on a regional scale to ensure reliable and accurate results. DL techniques are easily applied to image datasets as well. Table 2 shows the abbreviations used for ML and DL techniques.

Table 2. Abbreviations for ML and DL techniques

Abbreviation	ML and DL Techniques
SVM	Support Vector Machine
ULR	Univariate Linear Regression
RF	Random Forest
DNN	Deep Neural Network
RCNN	Region-Based CNN
DLMLP	Deep Learning Multi-Layer Perceptron
ANFIS	Adaptive-Network-Based Fuzzy Inference System
MLR	Multiple Linear Regression
PCR	Principle Component Regression
FIS	Subtractive Clustering Based Fuzzy Inference System
RBF NN	Radial Basis Function Neural Network
SVR	Support Vector Regression
PLSR	Partial Least Squares Regression
GBRT	Gradient Boost Regression Tree
RFR	Random Forest Regression
MLP	Multilayer Perceptron
DT	Decision Tree
DBSCAN	Density-Based Spatial Clustering of Applications With Noise
CNN	Convolutional Neural Network
BDT	Bagging Decision Tree
SMLP	Stepwise Multiple Linear Regression
BPNN	Back Propagation Neural Network

3.1 Classification of ML and DL Algorithms

The two primary categories of Machine Learning are classified as follows:-

- *Supervised Learning:* This type of machine learning involves training models with labelled data. The result is labelled data that is previously known.
- *Unsupervised Learning:* These Models are trained using unlabeled datasets in machine learning.

Some common ML algorithms used in soil nutrient analysis are [30]:-

1. *Random Forest (RF):* RF is a versatile algorithm that can be used for both classification and regression tasks and consists of multiple decision trees to make predictions.
2. *Support Vector Machine (SVM):* SVM is primarily utilized for classification purposes, but it can also be applied to regression tasks on occasion. It creates a hyperplane between different types of data.

3. *Extreme Gradient Boost (XGBoost):* XGBoost is an ensemble-based Machine Learning technique that utilizes the collective predictions of numerous individual decision trees to yield a more precise overall prediction.
4. *Multiple Linear Regression (MLR):* Multiple linear regression, which is a fundamental regression method, models the linear association between a solitary continuous dependent variable and multiple independent variables.

Some common DL algorithms used in soil nutrient analysis are [21]:-

1. *Convolutional Neural Network (CNN):* It is primarily utilized for examining structured data arrays, with a particular focus on image analysis. CNNs excel at detecting design features such as lines, gradients, circles, and even complex objects like eyes and faces in input images.
2. *Recurrent Neural Network (RNN):* It retains all computation-related information within its "memory." RNNs execute identical operations on all inputs or hidden layers to generate the output by employing the same parameters for each input, thereby minimizing the parameter set's complexity compared to other neural networks.
3. *Artificial Neural Network (ANN):* It can be trained to identify patterns and correlations in data, making them applicable to a broad spectrum of tasks, such as speech and audio recognition, natural language processing (NLP), and decision-making.

4   Different Types of Soil Dataset

This section addresses RQ1. There are various types of Soil data on which ML and DL are applied, which are categorized as follows:-

– *Image data* - Images are captured through various cameras, smartphones, UAVs, and satellites [1]. The images can be in RGB, multispectral, and hyperspectral forms. Mainly DL models are applied to the image dataset.
– *Chemical based data:* Soil nutrients are related to each other. They have correlated with each other, as phosphorus solubility decreases in higher soil pH soils where it reacts with soil calcium [50]. In this data type, input parameters are some soil nutrients used to predict other soil nutrients.
– *Hyperspectral data:* This type of data is collected with the help of UAVs and satellites and requires high processing and robust models for data analysis [18].

5   Review of ML and DL Techniques to Estimate Soil Nutrients

This section addresses RQ2 and RQ3. The search method in the introduction is used to find works suggesting ML and DL soil nutrient analysis approaches. Figure 2 shows the number of studies published annually from (2019-2022) on soil nutrient analysis using artificial intelligence.

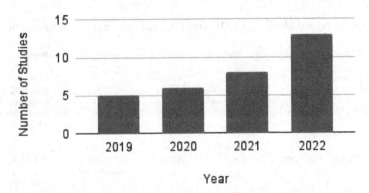

Fig. 2. Number of studies published per year on soil nutrient analysis using artificial intelligence

Table 3. Papers that use data ML and DL models to analyze macronutrients

Previous Work	Type of dataset	ML/DL	Nutrients	Location	Algorithm	Results
Liu et al. (2022). [17]	Spectral	ML, DL	Nitrogen	China	SVM, BPNN, PLSR	R2 value of SVM - 0.676, BPNN - 0.560, PLSR - 0.374
Yi et al. (2020). [47]	RGB	DL	Nitrogen, Phosphorous, Potassium, Calcium	Germany	CNN	Accuracy score is 98.4%
Farwa et al. (2020). [9]	Chemical	ML	Nitrogen, Potassium, Phosphorus	Pakistan	Linear, Ridge, Lasso, Elastic net and Bayesian Regression model	R2 value for Linear-0.057, Ridge-0.475, Lasso-0.304, Elastic net-0.258, Bayesian-0.113
Jin et al. (2020). [13]	Spectral	ML	Potassium	China	PLSR, SVR, GBRT, AdaBoost, Elastic net, Lasso, Ridge	Best R2 value is of AdaBoost and GBRT - 0.99
Peng et al.(2019). [25]	Spectral	ML, DL	Nitrogen, Potassium, Phosphorous	China	PLSR, BPNN and Genetic Algorithm (GA)-BPNN	Best RMSE value is for GA-BPNN that is 20.1%, 16.5% and 47.1% for nitrogen, potassium and phosphorous

Tables 3, 4, 5, 6 and 7 compare various ML and DL techniques for predicting soil macronutrients, micronutrients, soil pH, soil organic carbon (SOC) and soil organic matter (SOM). Each item is categorized in these tables according to the dataset type explained in Sect. 4. The tables include a thorough list of authors, their individual models, and the associated results. Table 8 illustrates the summary of papers that use AI models for multiple soil nutrients.

In recent years, AI has shown significant promising results in soil nutrient analysis. The study suggests that these techniques have the potential to accurately predict soil nutrient levels and identify important trends and patterns in soil composition.

Table 4. Papers that use data ML and DL models to analyze micronutrients

Previous Work	Type of dataset	ML/DL	Nutrients	Location	Algorithm	Results
Keshavarzi et al. (2022) [15]	Hyperspectral	ML	iron, manganese, zinc and copper	Northeast Iran	RF	RMSE values are 91%, 94%, 91% and 108% for iron, manganese, zinc and copper
Zhang et al. (2022) [49]	Hyperspectral	ML	Chromium and zinc	China	Deep Forest (DF21)	R2 - 0.58
Xu et al. (2022) [45]	Hyperspectral	ML	Iron	China	RF, XGBoost, CatBoost and SVR	SVR shows better RMSE value that is 2.465
Shi et al. (2021) [31]	Hyperspectral	ML	Zinc	China	RF	R2 value is 0.68
Wang et al. (2021) [40]	Hyperspectral	DL	Soil chromium	Eastern Junggar Coalfield in Xinjiang	Deep Neural Network	Accuracy score is 96.25%

Table 5. Papers that use data ML and DL models to analyze Soil pH

Previous Work	Type of dataset	ML/DL	Location	Algorithm	Results
Guo et al. (2022) [12]	Chemical	ML	China	SVM	R2 is 0.68
Natarajan et al. (2022) [19]	Spectral	ML	Tamilnadu	Gradient Boost Regression	RMSE - 0.92
Tejaswani et al. (2021) [41]	RGB	ML	India	MLP	Accuracy is 95%
Sunori et al. (2021) [34]	Chemical	ML	Uttarakhand	SVM - Linear, Quadritic and Cubic	MSE of Linear SVM-0.44, Quadritic SVM -0.43, Cubic SVM =0.48
Barman et al. (2019). [2]	RGB	ML, DL	Assam, India	Linear, ANN and KNN Regression	Regression coefficient of Linear Regression - 0.859, ANN - 0.940, KNN Regression - 0.893

Table 6. Papers that use data ML and DL models to analyze Soil organic carbon

Previous Work	Type of dataset	ML/DL	Location	Algorithm	Results
Tripathi et al. (2022) [37]	Spectral	DL	Rupnagar District, Punjab	DLMLP	MAE - 0.98, RMSE - 1.24, R2 - 0.684
Odebiri et al. (2022) [22]	Chemical	DL	South Africa	Deep Neural Network	R2 - 0.685
Yang et al. (2022) [46]	Satellite	DL	Anhui province, China	CNN	RMSE - 5.57% , R2 - 31.29%
Zadeh et al. (2020) [48]	Chemical	ML	Western Iran	RF, KNN, SVM, XGBoost	The performance metrics of RMSE-0.35% and R2-0.60 indicated that RF had the best results.
Nawar et al. (2019) [20]	Spectral	ML	Germany	RF	R2 - 0.80

Table 3 shows that mainly nitrogen, potassium, and phosphorus are analyzed by ML and DL algorithms. There is a need to investigate other soil macronutrients such as sulfur, calcium, and magnesium. The findings of Tables 4 and 5 present that mainly ML models are applied to predict soil micronutrients and soil pH, and more use of DL models is also required. In previous studies, authors utilized hyperspectral data for analyzing soil micronutrients, but there still needs to be more exploration of other soil data, such as chemical and RGB image data. Table 6 shows that researchers have used both ML and DL models to analyze soil organic carbon on different soil datasets. In contrast, Table 7 concludes that smartphone-based image datasets mainly predict SOM and chemical data, and hyperspectral data still need to explore. Finally, Table 8 concludes that researchers mainly predict macronutrients and mention SOM the least in their studies. The capacity of regression algorithms and neural networks to forecast continuous values based on input data makes them popular tools for soil nutrient analysis. Regression techniques, such as linear and support vector regression, describe the relationship between input variables and output values. Neural networks, on the other hand, can be used to analyse soil nutrients to discover patterns and connections between various soil characteristics and nutrient

Table 7. Papers that use data ML and DL models to analyze Soil organic matter

Previous Work	Type of dataset	ML/DL	Location	Algorithm	Results
Gorthi et al. (2021) [11]	RGB	ML	Different agroclimatic zones of West Bengal	RFR, SVR , AdaBoost Regression , Ridge	R2 - 0.88, RMSE - 0.28%
Taneja et al. (2021) [35]	RGB	ML, DL	Canada	SVM, DT, Linear Regression, Gausian Regression, Cubist Regression and ANN	ANN and Cubist Regression well with R2 values 0.91 and 0.72
Dong et al. (2021) [6]	Spectral	ML, DL	China	Cubist method, PLSR, SVM, ANN	Cubist method performed well among all with R2 value o.86
Fu et al. (2020) [10]	RGB	ML	Macdonald Campus Farm, McGill University, Quebec, Canada	ULR, SMLP	Below soil moisture content R2 is 0.936 and above soil moisture content R2 is 0.819 and without including soil moisture content R2 is 0.741
Chen et al. (2019) [4]	Chemical	ML	China	DT, BDT, RF and GBRT	RF and DT performed well with R2 value is 0.61

Table 8. Summary of the papers that use data ML and DL models to multiple soil nutrients [N1]-Macronutrients, [N2]-Micronutrients, [N3]-Soil pH, [N4]-Soil Organic Carbon (SOC), [N5]-Soil Organic Matter (SOM)

Previous Work	Type of dataset	ML/DL	Nutrient Type	Nutrients	Location	Algorithm	Results
Wei et al. (2022). [43]	Spectral	ML	[N1],[N3],[N5]	SOM, soil pH, Potassium, Phosphorous, Calcium, Magnesium	Luiz de Queiroz College of Agriculture at the University of São Paulo	MLR, PCR, lasso method	lasso method and PCR overperformed MLR approach with R^2 values range from 0.33-.96 and 0.03-0.84 for Vis-NIR and XRF sensor data
Wilhelm et al. (2022). [44]	Chemical	ML	[N1],[N2],[N3], [N4],[N5]	SOM, SOC, soil pH, Potassium, Phosphorous, Zinc, Iron, Manganese and Magnesium	Farmlands across the USA and Canada	RF and SVM	R2 for SVM is 0.056 and RF - 0.065
Pillai et al. (2022). [26]	Chemical	ML	[N1],[N2]	Sodium, Zinc, Potassium, and Copper	Southeastern United States	RF, XGBoost, Multi-layer Perceptron, Stacked autoencoder, Generative adversarial network	AUC score is 0.921
Wang et al. (2021). [39]	Chemical	ML	[N1],[N5]	SOM, Nitrogen, Potassium, Phosphorus	Hongliulin Coalfield on the Loess Plateau of China	DBSCAN clustering	The training set and testing set achieved accuracy scores of 98% and 95%, respectively.
Suchitra et al. (2020). [33]	Chemical	ML	[N1],[N2],[N3], [N4]	Phosphorous, Potassium, SOC, Boron and soil pH	North Central Laterites Agro-Ecological Unit (AEU), Kerala	Extreme Learning Machine Parameters	Accuracy score of Phosphorus-90%, Boron - 85%, SOC - 80%, Potassium - 78%, soil pH - 89%.
Gutierrez et al. (2022). [8]	Chemical	DL	[N1],[N2],[N3], [N4]	soil pH, SOC, Phosphorous, Potassium, Iron, Zinc, Boron, Chlorine, Copper and Manganese.	Soil samples were collected by individual farmers by the soil testing laboratory	Ensemble Deep learning techniques (ISNpHC-WVE)	Accuracy for Soil Nutrients - 0.9281 and soil pH - 0.9497
Padarian et al. (2019). [23]	Spectral	DL	[N1],[N3],[N4]	SOC, soil pH, and Nitrogen	LUCAS soil database, Europe	CNN	Compared with PLS regression and Cubist regression tree CNN decreased the error 87% and 62%

concentrations, which can subsequently be applied to forecast nutrient concentrations in soil samples. Since all soil nutrients are essential and contribute to plant growth, predicting every soil nutrient is vital.

Table 9. Open datasets of soil domain.

References	Dataset	Description	Research Focuses
Tavares et al. (2022) [36]	Spectral data of tropical soils using dry-chemistry techniques (VNIR, XRF, and LIBS): A dataset for soil fertility prediction	In addition to characterizing key soil fertility attributes (clay, organic matter, cation exchange capacity, pH, base saturation, and exchangeable P, K, Ca, and Mg) of 102 soil samples from a Brazilian agricultural area, the shared dataset also includes spectral data from VNIR, XRF, and LIBS sensors	Wei et al. (2022) [43] used this dataset for soil attribute prediction. Applied MLR, PCR and lasso method with R2 values ranges from 0.02-0.84
Riese et al. (2018). [27]	Hyperspectral benchmark dataset on soil moisture	In May 2017 in Karlsruhe, Germany, a five-day field campaign was used to collect the data for this dataset. A hyperspectral snapshot camera captures one hundred twenty-five spectral bands in 50 by 50-pixel images	Keller et al. (2018) [14] used this dataset to estimate soil moisture in which an extremely random tree model without preprocessing offers the best performance
Bolanos et al. (2023). [28]	Determination of phosphorus in the soil through the analysis of hyperspectral images	A hyperspectral cube was created for each of the 152 soil samples, consisting of 145 images in the VIS-NIR bands from 420 to 1000 nm. The data was collected in Columbia	No previous work has been done till now.
Tziachris et al. (2022). [38]	Soil data Grevena	The data is collected from northern Greece, Grevena. This dataset contains studies of soil parameters including pH, OM, EC, salinity, major elements (N, P, K, Mg), and several microelements (Fe, Zn, Mn, Cu, B) that have a big impact on plant nutrition	No previous work has been done till now

6   Open Datasets of Soil

The ability to assess trends and hidden patterns and make judgments based on the dataset is enabled by sufficient data quaThe ntities. Table 9 shows some open datasets of the soil domain. All these datasets are publicly available. Anyone can download these datasets and perform ML and DL models to analyze the soil nutrients.

7   Open Issues of ML and DL in Soil Research

This section addresses RQ4. The development of prediction methods like AI and ML is leaning toward the research area. Some of the following issues are analyzed after the literature survey:-

1. *Regional differences in soil:* It cannot be easy to create precise and trust-worthy AI models that can generalise successfully to new regions due to the variation in soil qualities across different geographical areas. This is due to the possibility of regional differences in the connections between soil nutrients and other environmental factors, including temperature, rainfall, and vegetation [32].
2. *Data impacts performance:* The computational complexity of deep learning algorithms is a critical factor affecting the training process's speed and efficiency. To improve the computational efficiency of deep learning algorithms, researchers and practitioners need to develop efficient optimization algorithms and reduce the complexity of the neural network [42].
3. *High-dimensional hyperspectral data:* The high dimensionality of hyperspectral data is due to its enormous number of spectral bands or channels. As a result, processing hyperspectral data can be difficult since it necessitates many calculations, which raises the cost of computing, increases processing time, and decreases accuracy [40].
4. *Natural light interference:* Due to the influence of natural light, taking soil samples with a smartphone camera for soil pH prediction can be complicated. The colour and appearance of the soil in the image can change depending on the intensity and direction of the sunlight and the presence of shadows, which can result in inaccurate pH predictions [2].

8   Conclusion

This review provides a comprehensive overview of the current knowledge in applying artificial intelligence to soil nutrient analysis. It draws attention to the deficiencies in the existing body of literature and pinpoints potential areas for further investigation. The review also explains different soil datasets like image, chemical, and hyperspectral data on which soil nutrient analysis is done. The review has also highlighted some open datasets for soil nutrient prediction. It provides a foundation for further research in the soil field and encourages continued exploration of the potential of AI for soil nutrient analysis. However, more research is needed to realize AI's potential in the soil field. Future research should focus on developing robust and scalable AI algorithms that can handle large and complex soil nutrient data, as well as the integration of AI with other technologies such as sensor networks and remote sensing.

References

1. Five ways satellite images, remote sensing and smartphones are combining to transform agriculture. https://www.cgiar.org/news-events/news/five-ways-satellite-images-remote-sensing-and-smartphones-are-combining-to-transform-agriculture/. Accessed 7 Feb 2023
2. Barman, U., Choudhury, R.D.: Prediction of soil pH using smartphone based digital image processing and prediction algorithm. J. Mech. Contin. Math. Sci. **14**, 226–249 (2019)
3. Chandraprabha, M., Dhanaraj, R.K.: Soil based prediction for crop yield using predictive analytics. In: 2021 3rd International Conference on Advances in Computing, Communication Control and Networking (ICAC3N), pp. 265–270. IEEE (2021)
4. Chen, D., et al.: Mapping dynamics of soil organic matter in croplands with MODIS data and machine learning algorithms. Sci. Total Environ. **669**, 844–855 (2019)
5. Diaz-Gonzalez, F.A., et al.: Machine learning and remote sensing techniques applied to estimate soil indicators-review. Ecolog. Indicat. **135**, 108517 (2022)
6. Dong, Z., Wang, N., Liu, J., Xie, J., Han, J.: Combination of machine learning and VIRS for predicting soil organic matter. J. Soils Sedim. **21**(7), 2578–2588 (2021). https://doi.org/10.1007/s11368-021-02977-0
7. Eli-Chukwu, N.C.: Applications of artificial intelligence in agriculture: a review. Eng. Technol. Appl. Sci. Res. **9**(4), 4377–4383 (2019)
8. Escorcia-Gutierrez, J., et al.: Intelligent agricultural modelling of soil nutrients and pH classification using ensemble deep learning techniques. Agriculture **12**(7), 977 (2022)
9. Farwa, U.E., et al.: Prediction of soil macronutrients using machine learning algorithm. Int. J. Comput. (IJC) **38**(1), 1–14 (2020)
10. Fu, Y., et al.: Predicting soil organic matter from cellular phone images under varying soil moisture. Geoderma **361**, 114020 (2020)
11. Gorthi, S., et al.: Soil organic matter prediction using smartphone-captured digital images: use of reflectance image and image perturbation. Biosyst. Eng. **209**, 154–169 (2021)
12. Guo, J., et al.: Mapping of soil pH based on SVM-RFE feature selection algorithm. Agronomy **12**(11), 2742 (2022)
13. Jin, X., et al.: Prediction of soil-available potassium content with visible near-infrared ray spectroscopy of different pretreatment transformations by the boosting algorithms. Appl. Sci. **10**(4), 1520 (2020)
14. Keller, S., et al.: Developing a machine learning framework for estimating soil moisture with VNIR hyperspectral data. arXiv preprint arXiv:1804.09046 (2018)
15. Keshavarzi, A., et al.: Spatial prediction of soil micronutrients using machine learning algorithms integrated with multiple digital covariates (2022)
16. Lamichhane, S., et al.: Digital soil mapping algorithms and covariates for soil organic carbon mapping and their implications: a review. Geoderma **352**, 395–413 (2019)
17. Liu, Z., et al.: Spatial prediction of total nitrogen in soil surface layer based on machine learning. Sustainability **14**(19), 11998 (2022)
18. Lu, B., et al.: Recent advances of hyperspectral imaging technology and applications in agriculture. Remote Sens. **12**(16), 2659 (2020)
19. Natarajan, V.A., et al.: Prediction of soil pH from remote sensing data using gradient boosted regression analysis. J. Pharm. Negat. Results **13**, 29–36 (2022)

20. Nawar, S., et al.: On-line vis-NIR spectroscopy prediction of soil organic carbon using machine learning. Soil Tillage Res. **190**, 120–127 (2019)
21. Odebiri, O., Mutanga, O., Odindi, J., Naicker, R., Masemola, C., Sibanda, M.: Deep learning approaches in remote sensing of soil organic carbon: a review of utility, challenges, and prospects. Environ. Monitor. Assess. **193**(12), 1–18 (2021). https://doi.org/10.1007/s10661-021-09561-6
22. Odebiri, O., et al.: Modelling soil organic carbon stock distribution across different land-uses in South Africa: a remote sensing and deep learning approach. ISPRS J. Photogramm. Remote Sens. **188**, 351–362 (2022)
23. Padarian, J., et al.: Using deep learning to predict soil properties from regional spectral data. Geoderma Reg. **16**, e00198 (2019)
24. Padarian, J., et al.: Machine learning and soil sciences: a review aided by machine learning tools. SOIL **6**(1), 35–52 (2020)
25. Peng, Y., et al.: Prediction of soil nutrient contents using visible and near-infrared reflectance spectroscopy. ISPRS Int. J. Geo-Inf. **8**(10), 437 (2019)
26. Pillai, N., et al.: An ensemble learning approach to identify pastured poultry farm practice variables and soil constituents that promote salmonella prevalence. Heliyon **8**(11), e11331 (2022)
27. Riese, F.M., Keller, S.: Hyperspectral benchmark dataset on soil moisture. In: Proceedings of the 2018 IEEE International Geoscience and Remote Sensing Symposium (IGARSS), Valencia, Spain, pp. 22–27 (2018)
28. Rivadeneira-Bola, F.E., et al.: Dataset for the determination of phosphorus in soil through the analysis of hyperspectral images. Data Brief **46**, 108789 (2023)
29. Shahare, Y., Gautam, V.: Soil nutrient assessment and crop estimation with machine learning method: a survey. In: Tavares, J.M.R.S., Dutta, P., Dutta, S., Samanta, D. (eds.) Cyber Intelligence and Information Retrieval. LNNS, vol. 291, pp. 253–266. Springer, Singapore (2022). https://doi.org/10.1007/978-981-16-4284-5_22
30. Sheeba, B., et al.: Machine learning algorithm for soil analysis and classification of micronutrients in IoT-enabled automated farms. J. Nanomater. **2022**, 5343965 (2022)
31. Shi, T., et al.: Digital mapping of Zinc in urban topsoil using multisource geospatial data and random forest. Sci. Total Environ. **792**, 148455 (2021)
32. Srivastava, P., Shukla, A., Bansal, A.: A comprehensive review on soil classification using deep learning and computer vision techniques. Multimedia Tools Appl. **80**(10), 14887–14914 (2021). https://doi.org/10.1007/s11042-021-10544-5
33. Suchithra, M.S., et al.: Improving the prediction accuracy of soil nutrient classification by optimizing extreme learning machine parameters. Inf. Process. Agricult. **7**(1), 72–82 (2020)
34. Sunori, S.K., et al.: Machine learning based prediction of soil pH. In: 2021 5th International Conference on Electronics, Communication and Aerospace Technology (ICECA), pp. 884–889. IEEE (2021)
35. Taneja, P., et al.: Multi-algorithm comparison to predict soil organic matter and soil moisture content from cell phone images. Geoderma **385**, 114863 (2021)
36. Tavares, T.R., et al.: Spectral data of tropical soils using dry-chemistry techniques (VNIR, XRF, and LIBS): a dataset for soil fertility prediction. Data Brief **41**, 108004 (2022)
37. Tripathi, A., et al.: A deep learning multi-layer perceptron and remote sensing approach for soil health based crop yield estimation. Int. J. Appl. Earth Observ. Geoinform. **113**, 102959 (2022)

38. Tziachris, P., et al.: Soil data Grevena. https://data.mendeley.com/datasets/r7tjn68rmw/1 (2022). https://doi.org/10.1016/j.dib.2022.108408
39. Wang, Z., et al.: Assessment of soil fertility degradation affected by mining disturbance and land use in a coalfield via machine learning. Ecolog. Indicators **125**, 107608 (2021)
40. Wang, Y., et al.: Hyperspectral monitor of soil chromium contaminant based on deep learning network model in the eastern Junggar coalfield. Spectrochimica Acta Part A Molecul. Biomolecul. Spectros. **257**, 119739 (2021)
41. Wani, T., Dhas, N., Sasane, S., Nikam, K., Abin, D.: Soil pH prediction using machine learning classifiers and color spaces. In: Joshi, A., Khosravy, M., Gupta, N. (eds.) Machine Learning for Predictive Analysis. LNNS, vol. 141, pp. 95–105. Springer, Singapore (2021). https://doi.org/10.1007/978-981-15-7106-0_10
42. Wankhede, D.S.: Analysis and prediction of soil nutrients pH,N,P,K for crop using machine learning classifier: a review. In: Raj, J.S. (ed.) ICMCSI 2020. EICC, pp. 111–121. Springer, Cham (2021). https://doi.org/10.1007/978-3-030-49795-8_10
43. Wei, M.C.F., et al.: Dimensionality reduction statistical models for soil attribute prediction based on raw spectral data. AI **3**(4), 809–819 (2022)
44. Wilhelm, R.C., et al.: Predicting measures of soil health using the microbiome and supervised machine learning. Soil Biol. Biochemis. **164**, 108472 (2022)
45. Xu, S.X., et al.: A comparison of machine learning algorithms for mapping soil iron parameters indicative of pedogenic processes by hyperspectral imaging of intact soil profiles. Eur. J. Soil Sci. **73**(1), e13204 (2022)
46. Yang, L., et al.: A deep learning method to predict soil organic carbon content at a regional scale using satellite-based phenology variables. Int. J. Appl. Earth Observ. Geoinform. **102**, 102428 (2021)
47. Yi, J., et al.: Deep learning for non-invasive diagnosis of nutrient deficiencies in sugar beet using RGB images. Sensors **20**(20), 5893 (2020)
48. Mahmoudzadeh, H., et al.: Spatial prediction of soil organic carbon using machine learning techniques in western Iran. Geoderma Reg. **21**, e00260 (2020)
49. Zhang, Z.H., et al.: On retrieving the chromium and zinc concentrations in the arable soil by the hyperspectral reflectance based on the deep forest. Ecol. Ind. **144**, 109440 (2022)
50. Zhao, J., et al.: Effect of annual variation in soil pH on available soil nutrients in pear orchards. Acta Ecol. Sinica **31**(4), 212–216 (2011)

IRPD: In-Field Radish Plant Dataset

Simrandeep Singh$^{(\boxtimes)}$ ⓘ, Davinder Singh ⓘ, Snigdha Agarwal ⓘ,
and Mukesh Saini ⓘ

Indian Institute of Technology Ropar, Rupnagar 140001, Punjab, India
{staff.simrandeep.singh, staff.davinder.singh,
staff.snigdha.agarwal,mukesh}@iitrpr.ac.in

Abstract. In this study, we present an in-field radish plant dataset
(IRPD) for the segmentation and counting of leaves. Precision agriculture
requires such a dataset to estimate the development, growth monitoring,
health status, and yield potential of plants using computer vision and
deep learning algorithms. The dataset consists of 6504 total images, out
of which 3252 were captured over ten weeks in an uncontrolled open
field environment, and the remaining 3252 were generated using image
augmentation. Images were taken from germination through harvesting
using different cameras. The total of 1025 images in the dataset are manually annotated for segmentation and leaf count. We present a baseline
for leaf count and segmentation using Detectron2 and UNet. We hope
that by making this dataset and annotated data available to the public,
we can encourage research in this field, where a lack of publicly available
in-field datasets is currently a barrier to advancement.

Keywords: Leaf count · dataset · segmentation

1 Introduction

The collection of image dataset, especially in the agriculture field, requires a
great deal of effort and workforce, which makes it expensive. Sharing the collected
dataset is therefore helpful for the researchers to access the data, which varies
in region, time, soil, and weather conditions [19].

From a phenotyping point of view, leaf segmentation and counting the number of leaves in plants is important for the developmental stage, plant growth
monitoring, flowering time, plant phenotyping, and yield estimation. However,
counting the number of leaves is a tedious task even when undertaken by experts,
as leaves frequently overlap each other [17]. It becomes more challenging when
rosette plants [18] are considered as shown in Fig. 1. All leaves must be precisely
taken into account by the automatic system to figure out the current growth
period and decide the best course of action.

This work was supported by DST, Government of India, for the Technology Innovation
Hub at the IIT Ropar in the National Mission on Interdisciplinary Cyber-Physical
Systems framework.

Fig. 1. Top and side views of rosette leaf structure of radish plant.

An in-field radish plant dataset (IRPD) is proposed in this paper. Radish is an annual or biennial cultivated vegetable plant in the mustard family. It is traditionally famous in many countries, and also cultivated and consumed with different names such as cultivar, Chinese radish, Japanese radish, or Oriental radish. Radish was portrayed on the Pyramids of Egypt, it may be one of the common agricultural plants in ancient Egypt. Radish and its uses were first mentioned nearly 2,000 years ago in China and about 1,000 years ago in Japan.

Its biological name is Raphanus sativus L. and is a member of the Crucifereae family. It has a short hairy stem and contains radical leaves. The radish fruit is available in different shapes based on the variety i.e. circular, long and conical, or tapered. Moreover, it has variety of colours such as white, yellow, pink, red, or purple, and also differs in size and weight from a few grams to kilogram. In Punjab, radish is usually grown on ridges to encourage healthy growth of plant. The height of ridge, plant to plant distance, and ridge to ridge distance is 25 cm, 45 cm, and 10 cm approximately as shown in Fig. 2.

The major contribution of this paper is that, (i) we are proposing a novel dataset of 6404 in-field radish plant images (IRPD) acquired using different cameras; (ii) data is acquired from an uncontrolled environment i.e.(in-field); (iii) images are manually annotated for leaves, number of leaves, age in days, and view information (side view or top view) by agriculture experts; and (iv) the collection period of the data is from germination till maturity.

The remainder of this article is organized as follows. Section 2 offers related work. Section 3 describes the dataset, while Sect. 4 discusses the baseline model implementation. Section 5 offers concluding remarks and future work.

2   Related Work

Public datasets are highly valued in the research community because they allow scientists to perform operations on the data and compare different algorithms with the state-of-the-art. Public dataset are well-established and play a significant role in many domains such as image processing, machine learning, time series, natural language processing, health care and image processing. Teimouri et al. [16] propose a dataset of fifty distinct tree species found in Sweden. A computer vision classifier is proposed using descriptors that classify leaves and

Fig. 2. Full view of field.

provide the name of the tree. Minervini et al. [13] provide a rigorous collection of image datasets, annotation, and procedure of annotation process for Arabidopsis and tobacco plants. To estimate the growth stage of a weed[16] proposes dataset of 9372 weed and crop coloured images. In the annotation process, leaf count for training dataset is provided.

Leaf Segmentation Challenge (LSC-2014) [13] was organized in CVPPP 2014, and a dataset having raw and annotated images was released. The same pattern followed in coming events, in which the Leaf Counting Challenge (LCC-2015) and (LCC-2017) were organized. Mortensen et al. [14] propose an oil radish growth dataset containing image and field data describing RGB images and weather data, but they have only labeled 95 images, and the rest, 5287 are unlabeled images. Fan et al. proposes both a multi-scale segmentation model and a regression model for leaf counting [7].

Leaf dataset proposed in the semi-controlled environment includes 'sugar beet and weed dataset' [4], 'organic carrot and weed dataset' [11] and 'white clover and grass species' [1]. Dobrescu [5] discusses different approaches for leaf count, such as counting via detection, segmentation, density estimation, and direct count. As reviewed above, the problem of leaf counting is dealt with two ways: (i) counting using leaf segmentation or detection: Instance or semantic segmentation is performed at the pixel, segment, and object levels using different CNN models [12,17]; (ii) considering the task as a regression problem: Counting leaves using a regression model is a bit easier as it only requires the total number of leaves in the training image [5,10].

Barbedo [3] demonstrated that even in the absence of a large dataset, deep learning models can be trained if a database has significant quality. Moreover, it may be artificially enhanced and diversified using different augmentation techniques.

There are many databases reported in the literature, but most of them are recorded in controlled environments. Very few databases are available that are taken directly from the agricultural field, i.e., under natural environments. This is the reason why the majority of experiments in the literature, conducted in a controlled environment show higher accuracy [2]. If the model is trained on the data of controlled environment and tested on real-time in-field data, Ferentinos [8] observed that there is a drop in accuracy from 99.5% to around 33%.

We are confident that other investigations will reveal similarly poor performance on real-world datasets. It occurs due to a lack or unavailability of real-time in-field labeled dataset, which is a time-consuming, labor-intensive, and expensive task.

3   Dataset and Problem Description

The leaf is the most important part of the plant, as it is responsible for photosynthesis and the generation of other nutrients. Growth stages, health status, and growth rate have a direct relationship with the number of leaves. Thus, the growth of any plant may be estimated based on the number of leaves [13].

We propose a dataset of in-field radish plants dataset (IRPD) collected in a natural environment with overlapped plant leaves. The dataset images are collected in fields across Ropar district in Punjab (India) (30.974922N/76.500364W).

3.1   Dataset Description

We have captured the data under uncontrolled conditions i.e. from the agricultural field directly, which greatly increases its research potential. The overall cycle of data acquisition, labeling, and evaluation is represented in Fig. 3. We highlight some salient features and characteristics of the data below.

Field Data: IRPD is collected in the uncontrolled and natural range of conditions, which makes it more comprehensive, diverse, complex, realistic, and informative. It provides relevant visual manifestations that accurately depict, what an operator would observe in the field. A model that has been trained on a collection of controlled condition images actually won't be able to generalize information from a more complex environment, including all the intricate details of leaf architecture. This type of dataset is most preferable for developing an operational autonomous intelligent system.

Data Annotation/Labeling: Manual annotation is carried out by agriculture experts at various growth phases, which is labor-intensive but less error-prone. More details can be seen in Table 2, which shows the distribution of images on

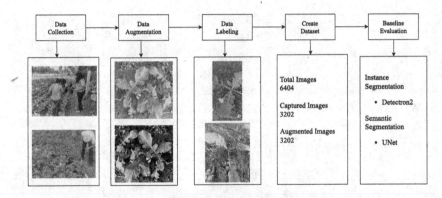

Fig. 3. Overall data acquisition cycle

Fig. 4. Field view demonstrating overlapping of plants.

a weekly basis. As the week progresses, the number of labeled images decreases; this is because when multiple plants are grown together in a field, the percentage of overlap between them increases, as shown in Fig. 4. After overlapping, it becomes very difficult or impossible to annotate images.

Duration: IRPD contains a set of ten-week images from the day of germination until maturity on a weekly basis. We can observe the week-wise growth of the plant and its life cycle, as shown in Fig. 5.

Quality: The dataset contains high-resolution (HD or 4K) images taken from both top and side views.

3.2 Field Setup and Acquisition Protocol

The reason for selecting the radish plant for this problem is the structure of the leaves. The lobed-shaped leaves of radish plants form a basal rosette, in which the ground level has horizontal and circular leaves. These oblong-shaped leaves arise from the top of the root and measure 5–30 cm (2–12 in.) in length. As the plant matures, the overlapping of leaves rises, and even the overlapping of different plants also increases.

Germination of radish usually starts in 5 to 6 days in the winter season, and the growth of radish plants becomes massive after five weeks. It becomes challenging to capture the images after that period. We have tried to capture

(a) Week 1   (b) Week 2   (c) Week 3   (d) Week 4   (e) Week 5

(f) Week 6   (g) Week 7   (h) Week 8   (i) Week 9   (j) Week 10

Fig. 5. Radish lifecyle

Table 1. Description of the camera system and acquisition parameters.

Acquisition Parameters	Value
Camera	Mobile camera of (Google Pixel 5, Onelus Note CE, and iphone 14 pro), RGB sensor of Fluke TiX580 Infrared Camera
Focal length	4.4 mm
Mean distance to ground	457 mm
Coordinates	30.974922N/76.500364W (232 mm)

images of plants from all possible viewing angles. The images are acquired at midday on a sunny day under in-field conditions at intervals of a week.

Throughout the data collection drive, the plant samples were fixed. To ensure an approximately uniform sampling from the field, sampling areas were preallocated. Corner ridges were selected for the sampling area, to avoid damaging the plants. Throughout the data collection drive, the same sampling area was under consideration, and the acquisition parameters are mentioned in Table 1. The data collection drive continued until maturity, i.e. for ten weeks. Before the first week of data collection, a stick with a tag was placed near the corner of each sampling area. When capturing images during the data collection drive in the following subsequent weeks, the stick was helpful in locating the sampling areas.

3.3 Dataset and Annotation Format

An online, open-source, and visual annotation tool "LabelMe" was used to annotate plant images. The leaves were marked using polygons, rectangles, circles, and points. In the labeling process, leaf shapes are individually annotated for segmentation by experts, as shown in Fig. 6 and associated data is provided for the learning process. IRPD can be used to train the system for per-leaf segmentation or direct image-to-count using a regression model.

Fig. 6. LabelMe Software used for data annotation.

As we know, the leaves of the radish plant grow in clusters and form a rosette structure. Thus, the overlapping of leaves makes this problem more complex and interesting. We were unable to label the leaves, where the overlapping is more than 80%. In such cases, the number of leaves counted by an expert is listed in the CSV file for reference.

To increase the count of data images, image augmentation techniques such as contrast and rotation are applied to the original dataset. Another 3252 images were created using such techniques.

Considering it to be the regression problem, the number of leaves can be estimated with the availability of only leaf count as ground truth.

3.4 Dataset Availability

The dataset proposed, along with the annotation data analyzed in this study, is readily available online for research purposes with a few terms and conditions. It can be downloaded from https://github.com/mriglab/IRPD (Fig. 7).

Fig. 7. In field data acquisition.

Table 2. In-field radish plants dataset (IRPD) distribution.

Week	Data Images	Augmented Images	Labelled	Unlabelled	Total Images
1	259	259	220	298	518
2	211	211	170	252	422
3	324	324	260	388	648
4	489	489	325	653	978
5	446	446	50	842	892
6	339	339	-	678	678
7	410	410	-	820	820
8	274	274	-	548	548
9	206	206	-	412	412
10	294	294	-	588	588

3.5 Potential Applications

We now list a few key application scenarios that would benefit from the availability of such a dataset.

Segmentation: IRPD can easily be used for training and testing of data for instance/semantic segmentation. In instance segmentation, different occurrences of the same class are segmented separately. This can be applied to multiple agriculture applications such as crop and weed classification, disease & pest detection, and leaf counting. It is well suited to segment the class of every pixel or different occurrences of the same class even for overlapping cases. In semantic segmentation, we can identify the class of every pixel in a given image but it doesn't differentiate among different instances belonging to the same class.

Classification: IRPD is appropriate for classification of plant, weed, soil, and other classes. It can also be used in plant recognition system like leaves, stem and soil, etc. Identification of plant species traditionally involves a lot of human labor. Pattern recognition and computer vision techniques are widely used in research. Our dataset is appropriate for training, since the plant components utilized in identifying different plant species, such as flowers and leaves, are accurately represented. It can also be used in plant part recognition like leaves, stem and soil, etc.

Growth Model: IRPD can help to develop a growth model, as we have images from every stage of growth. We can use this dataset to calculate the maturity time, average growth rate or crop's present growth stage. Growth is a key factor in determining the health and production of crops. It also serves as an important mediator of competitive interactions in plant communities. Thus, it is crucial for plant evolution, ecology, and agriculture science to understand what restricts plant development. This dataset can help develop a growth model of the crop as we have images from every stage of growth. We can use this dataset to calculate the necessary time and the average crop growth rate. It can also be used to create models that identify the crop's present growth stage.

Diseases Detection: IRPD contains a collection of pest images from which an automated diagnosis of pests and illnesses can be determined using deep learning, computer vision, and pattern recognition. For farmers, a key concern is the automated diagnosis of pests and illnesses that impact crops. Conventional approaches to computer vision and pattern recognition have limits when dealing with such difficult issues. But in recent years, there has been a rise in interest in deep learning, notably in the detection and recognition of biotic stressors from in-field images of plants. Our dataset is, therefore, ideal for tackling these issues. Since the images were taken using cellphones, they were impacted by changes in lighting, complicated backgrounds, image noise, and other factors.

Yield Estimation: IRPD provides an opportunity for researchers to test their yield-estimation algorithms in realistic scenarios. Estimating crop yields is a crucial part of crop management. The present manual yield estimating method requires a lot of time, effort, and is unreliable. Researchers working on computer vision-based systems for automated, quick, and accurate yield estimation to address this problem. The dataset provides an opportunity for the researchers to test their yield-estimation algorithms in realistic scenarios.

4 Baseline Evaluation

To provide baseline models for this challenge, we trained our dataset on two network models. Instance segmentation and semantic segmentation tasks are addressed using Detectron2 [15] and UNet [6] respectively. In instance segmentation, we detect the leaves in the image and count the number of leaves. In this experiment, the Detectron2 library is used to train the labeled images. We used the Faster RCNN model from the Detectron 2's model [9], and the experiments were performed using 50 epochs, batch size equal to 1, a learning rate of 0.005, and a confidence rate of 0.7. Second, we used UNet [6] for semantic segmentation to identify different classes of elements present. For UNet, the experiments were performed using 20 epochs with batch size 3 and a learning rate of 0.005. In the image, each pixel is classified as either radish, grass, weed, soil, equipment, or stubble. Pixels labeled as unknown in manual annotations are ignored during evaluation and can thus be ignored during training (Fig. 8).

The results obtained by the Faster R-CNN model implemented with the Detectron2 library using ResNet50 and ResNet101 for instance segmentation tasks are presented in Table 3. It represents that the Detectron2 Faster RCNN implementation, which uses ResNet50 as its backbone, performs better than ResNet101, with $AR_{max=100} = 87.7\%$ as opposed to $AR_{max=100} = 82\%$.

This suggests that the Detectron2 with ResNet50 as backbone implementation was more successful in recovering the proper instances. Since the object of interest appears more than once in every image, it can also be seen that $AR_{max=1}$ for the two solutions was low. The results obtained by UNet for semantic segmentation can be seen in Fig. 5 (Fig. 9).

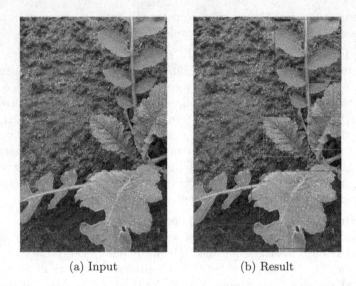

(a) Input   (b) Result

Fig. 8. Detectron2 Results

Table 3. Results obtained by Detectron2 Faster RCNN.

Parameters	ResNet50	ResNet101
AP	86.0	78.82
$AP_{IoU=0.5}$	80.386	96.87
$AP_{IoU=0.75}$	100	96.87
$AR_{max=1}$	40	39.1
$AR_{max=10}$	87.7	82.0
$AR_{max=100}$	87.7	82.0

(a) Input   (b) Result

Fig. 9. Semantic Segmentation results using UNet

UNet helps in semantic segmentation of the image which can be utilized in multiple problems. We can measure the performance of UNet based on AIoU and AC metrics in Table 4.

Table 4. Results obtained by UNet.

Model	AIoU	AC
UNet	86.43	87.26

5   Conclusion

In this paper, a dataset of radish plants (IRPD) collected from the field using various mobile phone cameras over the ten week period is proposed. The dataset is available for download and use for academic and research purposes. Radish plant is having rosette leaf structure, so counting the number of leaves and segmenting the leaves is very challenging job. Annotation of leaves is carried out in Labelme software, When overlapping of leaves becomes more than eighty percent then only total leaf count per plant is provided as annotation. IRPD is analyzed for instance and semantic segmentation using Detectron2 and UNet model. This data is collected for one season only, in the future work we would like to extend this dataset for different seasons with different field locations.

References

1. Bakken, A.K., Bonesmo, H., Pedersen, B.: Spatial and temporal abundance of interacting populations of white clover and grass species as assessed by image analyses. Dataset Papers Sci. (2015) (2015)
2. Barbedo, J.G.A.: Impact of dataset size and variety on the effectiveness of deep learning and transfer learning for plant disease classification. Comput. Electron. Agric. **153**, 46–53 (2018)
3. Barbedo, J.G.A.: Plant disease identification from individual lesions and spots using deep learning. Biosys. Eng. **180**, 96–107 (2019)
4. Chebrolu, N., Lottes, P., Schaefer, A., Winterhalter, W., Burgard, W., Stachniss, C.: Agricultural robot dataset for plant classification, localization and mapping on sugar beet fields. Int. J. Robot. Res. **36**(10), 1045–1052 (2017)
5. Dobrescu, A., Valerio Giuffrida, M., Tsaftaris, S.A.: Leveraging multiple datasets for deep leaf counting. In: Proceedings of the IEEE International Conference on Computer Vision Workshops, pp. 2072–2079 (2017)
6. Enshaei, N., Ahmad, S., Naderkhani, F.: Automated detection of textured-surface defects using unet-based semantic segmentation network. In: 2020 IEEE International Conference on Prognostics and Health Management (ICPHM), pp. 1–5. IEEE (2020)
7. Fan, X., Zhou, R., Tjahjadi, T., Das Choudhury, S., et al.: A segmentation-guided deep learning framework for leaf counting. Front. Plant Sci. 1466 (2022)

8. Ferentinos, K.P.: Deep learning models for plant disease detection and diagnosis. Comput. Electron. Agric. **145**, 311–318 (2018)

9. Girshick, R.: Fast R-CNN. In: Proceedings of the IEEE International Conference on Computer Vision, pp. 1440–1448 (2015)

10. Giuffrida, M.V., Minervini, M., Tsaftaris, S.A.: Learning to count leaves in rosette plants (2016)

11. Haug, S., Ostermann, J.: A crop/weed field image dataset for the evaluation of computer vision based precision agriculture tasks. In: Agapito, L., Bronstein, M.M., Rother, C. (eds.) ECCV 2014. LNCS, vol. 8928, pp. 105–116. Springer, Cham (2015). https://doi.org/10.1007/978-3-319-16220-1_8

12. Kumar, J.P., Domnic, S.: Image based leaf segmentation and counting in rosette plants. Inf. Process. Agric. **6**(2), 233–246 (2019)

13. Minervini, M., Fischbach, A., Scharr, H., Tsaftaris, S.A.: Finely-grained annotated datasets for image-based plant phenotyping. Pattern Recogn. Lett. **81**, 80–89 (2016)

14. Mortensen, A.K., Skovsen, S., Karstoft, H., Gislum, R.: The oil radish growth dataset for semantic segmentation and yield estimation. In: 2019 IEEE/CVF Conference on Computer Vision and Pattern Recognition Workshops (CVPRW), pp. 2703–2710. IEEE (2019)

15. Pham, V., Pham, C., Dang, T.: Road damage detection and classification with detectron2 and faster R-CNN. In: 2020 IEEE International Conference on Big Data (Big Data). pp. 5592–5601. IEEE (2020)

16. Teimouri, N., Dyrmann, M., Nielsen, P.R., Mathiassen, S.K., Somerville, G.J., Jørgensen, R.N.: Weed growth stage estimator using deep convolutional neural networks. Sensors **18**(5), 1580 (2018)

17. Xu, L., Li, Y., Sun, Y., Song, L., Jin, S.: Leaf instance segmentation and counting based on deep object detection and segmentation networks. In: 2018 Joint 10th International Conference on Soft Computing and Intelligent Systems (SCIS) and 19th International Symposium on Advanced Intelligent Systems (ISIS), pp. 180–185. IEEE (2018)

18. Zhang, Q., Liu, Y., Gong, C., Chen, Y., Yu, H.: Applications of deep learning for dense scenes analysis in agriculture: a review. Sensors **20**(5), 1520 (2020)

19. Zheng, Y.Y., Kong, J.L., Jin, X.B., Wang, X.Y., Su, T.L., Zuo, M.: Cropdeep: the crop vision dataset for deep-learning-based classification and detection in precision agriculture. Sensors **19**(5), 1058 (2019)

Fast Rotated Bounding Box Annotations
for Object Detection

Minhajul Arifin Badhon$^{(\boxtimes)}$ ⓘ and Ian Stavness ⓘ

Computer Science, University of Saskatchewan, Saskatoon, SK S7N 5C9, Canada
minhajul.arifin.badhon@gmail.com, ian.stavness@usask.ca

Abstract. Traditionally, object detection models use axis-aligned bounding boxes (AABBs), but these are often a poor fit for elongated object instances, such as wheat heads. Using rotated bounding box (RBB) annotations can improve object detection results, but RBBs are much more time-consuming and tedious to annotate than AABBs for large datasets. In this work, we propose a novel annotation tool for producing high-quality RBB annotations with low time and effort. The tool generates accurate RBB proposals for objects of interest as the annotator makes progress through the dataset. It can also adapt available AABBs to generate RBB proposals. Furthermore, a multipoint box drawing system is provided to reduce manual annotation time. Across three diverse datasets, we show that the proposal generation methods can achieve a maximum of 88.92% manual workload reduction. We also show that our proposed manual annotation method is twice as fast as the existing system with the same accuracy. Lastly, we publish the RBB annotations for two public datasets.

Keywords: Computer vision · Object detection · Aerial imaging · Wheat

1 Introduction

Object detection is an important and well-studied computer vision task. With the advent of large scale benchmark datasets [13,19], deep learning based detection algorithms have advanced rapidly [25]. However, these datasets primarily contain natural images labeled with axis-aligned bounding boxes (AABB), which do not localize elongated and rotated objects well. This is particularly problematic in aerial image datasets, due to the immense scale variations as well as random rotations of the objects therein (e.g., plants, ships). Also, AABBs fail to isolate the objects properly in crowded scenes, such as images of plants and crops, as adjacent bounding boxes heavily overlap each other (Fig. 1). The inclusion of background features or adjacent objects within an AABB potentially degrades detection performance and precise localization. To overcome this issue, object annotation with rotated bounding boxes (RBB) has been explored for detecting ships [20], vehicles [28], text [32] and other categories of objects [12].

© The Author(s), under exclusive license to Springer Nature Switzerland AG 2023
M. K. Saini et al. (Eds.): ICA 2023, CCIS 1866, pp. 99–115, 2023.
https://doi.org/10.1007/978-3-031-43605-5_8

Most previous RBB-based object detection methods are based on supervised learning and require a large amount of annotated images with RBBs around objects for training. Creating RBB annotations is more tedious and time-consuming than annotating AABBs with current tools because it requires iteratively rotating and resizing bounding boxes until they are a good fit. Annotation tools that support directly drawing RBBs are not widely available. In this work, we aim to fill this gap by integrating new RBB annotation features to the existing open-source annotation tool LabelImg [1,2]. The modified LabelImg tool is publicly available at https://github.com/p2irc/FastRoLabelImg to enable annotators to efficiently label large image datasets with RBBs.

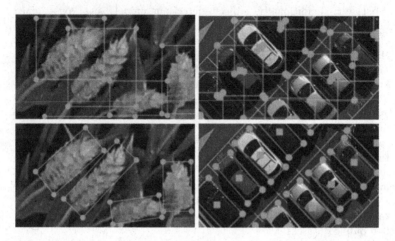

Fig. 1. Adjacent axis-aligned bounding boxes (AABB, top) have higher overlap compared to rotated boxes (RBB, bottom).

In order to reduce annotation workload, our tool automatically provides RBB proposals assisted by computer vision techniques. We use pre-trained models as well as an iteratively trained object detector. The latter model incrementally learns about the dataset as new annotations are available. As the annotator goes through the images in the dataset, the model is continually and seamlessly retrained and provides improved RBB proposals. We also explore ways in which *a priori* AABB annotations can be used to improve the accuracy of RBB proposals, for datasets with existing AABB annotations.

We evaluate the effectiveness of our developed methods using images drawn from the GWHD dataset [11], CARPK dataset [16], and Airbus ship detection challenge dataset [3]. The selected images (~1000) of the GWHD and CARPK dataset are manually annotated using RBBs and publicly made available at https://doi.org/10.6084/m9.figshare.13014230.v1 to encourage further research in this domain. The proposed RBB proposal generation methods attain a minimum workload reduction of 65.80% for the wheat dataset and a maximum of 88.92% for the ship dataset.

In addition to providing RBB proposals, our tool also includes a multipoint box drawing mechanism to speed-up manual bounding box creation. This system simplifies the process of drawing RBBs which is otherwise quite cumbersome in existing tools. We evaluate our proposed RBB drawing interface through a participant study and show that our two proposed variants are significantly faster than the traditional method and achieve the same level of annotation accuracy.

The contributions of this paper include: 1) development of multiple methods based on detection models and image processing to automatically generate RBB proposals for specific object classes; 2) design of a multipoint drawing system to accelerate the process of manually creating RBB annotations; 3) integration of existing AABB annotations, if available, to improve RBB proposals; 4) evaluation of the proposed RBB generation methods on subsets of three publicly available datasets (GWHD dataset, CARPK dataset, Airbus ship dataset); 5) assessment of the effectiveness of the presented RBB drawing system by conducting a participant study; and 6) publicly-available RBB annotations for subsets of two datasets (GWHD dataset, CARPK dataset).

2   Related Works

Bounding Box Annotation Tools. For automated content analysis, fully supervised computer vision models rely on manual annotation of images. To facilitate manual annotation, a lot of annotation tools have been developed over the years which allow different shapes such as box, polygon, circle, point, etc. to describe the object of interest. Label Studio [29] and a modified version of LabelImg [2] support drawing of an RBB in three steps. First, an AABB is drawn which is followed by rotating the box in any direction using keypress or mouse drag to match the orientation of the encapsulated object. Finally, the width and height of the box are adjusted again to fit the object. This process is labor-intensive and time-consuming when thousand of object instances need to be annotated. Some of the tools [4,5] allow pre-trained machine learning models to automate the annotation process but are limited by the number of classes the pre-trained model supports.

Weakly Supervised Object Localization. In weakly supervised object localization approaches image-level labels that describe the classes of objects present in the image are used to train object detectors instead of actual bounding boxes. Though these techniques [14,33] eliminate the need for bounding box annotations, the trained object detectors are significantly weak in terms of accuracy. Fully supervised object detectors with AABBs have twice the accuracy than weak detectors. Alternative forms of cheap human supervision have also been explored in recent works. In the box-verification series [22], human verification signals are used in both retraining an object detector and re-localizing the objects in the images via proposals iteratively. The trained detector achieves better object localization compared to WSOL approaches as only the correct bounding boxes are used in the training. In [18], the researchers proposed an Intelligent

Annotation Dialogues (IAD) agent that can determine the most time-efficient sequences of annotation actions to produce bounding box annotations for an image when image-level labels are available. It demonstrates that initially, the majority of the bounding boxes need to be manually drawn. But, as the detector gets stronger, box verification grows in number. Among others, eye-tracking [21] has been utilized to train object detectors, and click supervision has been leveraged for obtaining high qualify object annotations [23] as well as learning semantic segmentation models [8].

Human-Machine Collaborative Annotation. Computer vision models pre-trained on large scale benchmark datasets [13,19] are not strong enough to detect all the object instances in complex scenes. This becomes more evident when the intention is to build a new dataset that contains uncommon classes of objects. To solve the challenging task of getting high-quality object annotations efficiently, various works have combined the responses of computer vision models with human input. Adhikari et al. [7] divided the dataset into two parts. Then, a Faster RCNN model was trained using the manually annotated first part of the dataset to generate AABB proposals on the second part. These proposals were further refined by a human annotator with actions like removing incorrect boxes and drawing new ones. Adhikari [6] built on this approach by iteratively training the model with small batches of labeled images and proposals were generated on the next batch in the line. The focus of this work was not on developing an annotation tool but to compare different ordering of images to train a strong detector as early as possible. Getting a better model in early iterations helps to reduce the manual annotation effort as well as the total time needed. Several other works used human-machine collaborative annotation effort to develop interactive object segmentation [10,17], interactive video annotation [30] and attribute-based classification [9,24] methods.

3   Method

In this work, we aim to produce RBB annotations for better localization instead of the traditional AABBs. Rotating and re-scaling AABBs manually is tedious, therefore we design a multipoint drawing system to easily create a RBB from scratch. Our tool can also adapt and filter existing AABB annotations, when available, to automatically propose RBBs and speed up the annotation process.

3.1   Manual RBB Annotation

The purpose of the multipoint drawing system is to facilitate the process of manual bounding box creation. There are three configurations available categorized based on the number of points needed to be drawn to create a bounding box. The 2-point system is the traditional annotation method for drawing AABBs. The other two configurations can be used to create RBB annotations.

3-Point System. In this approach, the annotator can draw three corner points to create a bounding box of any rotations. Following the Fig. 2a, if the drawn points are A, B, and C' in this order, then first we calculate the perpendicular distance d from C' to the line AB. Afterward, we determine the coordinates of point C at a perpendicular distance d from B such that the point C lies on the same side of AB as C'. We take these additional steps to make sure that the line BC is perpendicular to AB through the point B. Given the corner points $A(x_1, y_1)$, $B(x_2, y_2)$ and $C(x_3, y_3)$ of the box, we can determine the other corner point $D(x_4, y_4)$ as: $x_4 = x_1 + x_3 - x_2$, and $y_4 = y_1 + y_3 - y_2$.

4-Point System. In the 4-point drawing system (Fig. 2b), if the points drawn are A, B, C', and D', we find the center O of the line AB. Then, similar to the 3-point drawing system, we consider points A, O, and D' to calculate the position of the point P. Consecutively, we calculate the coordinates of Q using A, O, and C'; the coordinates of R using B, O, and C'; the coordinates of S using B, O, and D'. Thus, we have all the coordinates of the RBB $PQRS$.

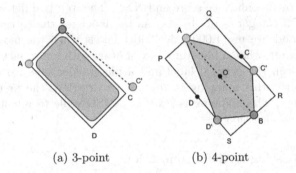

(a) 3-point   (b) 4-point

Fig. 2. Multipoint box drawing system for objects of interest (grey shapes). The colored dots are clicked around the objects in the order: Red $\Rightarrow$ Green $\Rightarrow$ Blue $\Rightarrow$ Orange. (Color figure online)

3.2 RBB Proposals

When no AABBs are available for an image, we use object detection models capable of predicting RBBs to generate proposals for the specified classes of objects in the image. We integrate two object detection models in the tool.

Pre-trained Detector. The pre-trained detector is based on the Faster R-CNN [26] framework embedded with ROI Transformer [12] to be able to predict the RBBs. The predicted bounding boxes can be denoted as $\{(x_i, y_i), i = 1,2,3,4\}$, where (x_i, y_i) denotes the positions of the box corners in the image. The pre-trained detector is trained on the DOTA [31] dataset which consists of

2,806 aerial images and the contained objects belong to 15 classes including ship, plane, storage tank, ground track field, tennis court, baseball diamond, basketball court, bridge, harbor, small vehicle, large vehicle, roundabout, helicopter, basketball court, and soccer ball field. So, this detector can support up to 15 categories when used to generate proposals. The predictions can also be filtered by confidence score and non-maximum suppression (NMS) before finalizing the proposals. The user can pre-configure the confidence score and overlap threshold for NMS through the interface of the tool.

Iterative Detector. We also use an object detector (Faster R-CNN with ROI Transformer) that is updated iteratively on previously annotated data using vertical re-training. The approach is motivated by batch-mode active learning [27]. In this tool, we use a batch size of 1. So, when a particular image is completely annotated by the user and annotations are saved, the newly available RBBs are utilized to retrain the detector right away. Each time the proposal generation step is requested, the most updated detector is used to generate the proposals for the selected categories. Similar to the pre-trained detector, the predictions can be filtered by a confidence score and NMS. The iterative detector can support up to 1,000 categories of objects. As the labels of the categories are not known beforehand, we use 1,000 placeholder labels when the model is configured initially. Later, when a bounding box appears with a new label, a mapping is created between the original label and a placeholder label to continue the training. Similarly, this mapping is also used to retrieve the predictions of a specified category. The mapping is saved as a JSON file to maintain its state among multiple annotation sessions.

3.3   Using Existing AABB Annotations

When AABBs are available, we leverage the bounded regions to apply segmentation based approaches to identify the pixels of interest and produce the RBB proposals with additional post-processing. However, we also use the object detection models to convert the AABBs to RBBs and if certain boxes are missed in this conversion, the tool falls back to the segmentation based approaches to fill the gap.

Filtering Proposals with IOU. If RBB proposals are requested for the available AABBs in an image using the object detection models, then we first get the predictions on the image for the specified categories of objects. Then, the predicted proposals are associated with the given AABBs using Intersection-over-Union (IoU) considering both of the boxes as polygons. This IOU-based matching process filters out many of the false positives. The IOU threshold for box matching can be set by the user in the tool.

Adapting AABBs with Thresholding. We make use of both traditional computer vision techniques as well as a deep learning-based instance segmentation model to serve the purpose of proposal generation from AABBs. Though a deep learning-based model can detect a large number of object classes, the detection is limited by the number of classes the model is pre-trained on. To overcome this issue, a set of assumptions is made about the objects inside the bounding boxes to develop an image processing based approach in support of the segmentation model. The image processing based approach might construct a better bounding box proposal than the model if the object of interest satisfies the assumptions.

We assume that the AABBs fit the objects as tightly as possible and that the centers of the bounding boxes are on the encapsulated objects. Also, we presume that the histograms of the pixel intensities in the grayscale patches formed by the bounding boxes will be bimodal.

Let's define p as a patch image bounded by an AABB b and i as an object instance of class c which is annotated by b. We convert the patch p to a grayscale image and determine a threshold value t using Otsu's method for the patch. We also calculate a mean pixel value m for the middle 10% region of the grayscale patch p. Then, m and t are used to determine the foreground and background pixels of the patch. If m is greater than t, then all the pixels having an intensity value greater than t are marked as foreground (in white color). Alternatively, if m is less than t, then all the pixels having an intensity value greater than t are marked as background (in black color).

We do a series of post-processing steps to get the final RBB proposal after getting the region of interest. Firstly, we might get one or multiple regions of interest. But, we know that a bounding box should contain one object at most. So, if multiple regions of interest are detected in a patch p, we try to find the best suitable region by weighting each of the regional areas with the distance between the center of the region and the center of the patch. Formally, if there are n regions, then we calculate the weighted area for the i_{th} region as $w_i = \text{area}(i)/\text{distance}(c_i, c_p)$ where $i = \{1 \ldots n\}$, c_i is the center of the i_{th} region and c_p is the center of the patch p. Afterward, the region with the maximum weighted area is chosen to apply the post-processing steps.

Subsequently, we find an ellipse that has the same second moments as the selected region of interest. The length of the minor axis of the ellipse is used as the height of the RBB. Furthermore, we extend the major axis on both sides and obtain the two intersecting points with the AABB. The distance between these points is used as the width of the RBB. As for the rotation of the box, the angle between the major axis of the ellipse and the positive x-axis is used. The range of the angle is between 0 to 180° in the pixel coordinate system (positive rotation is clockwise from the x-axis). Lastly, the center of the AABB is considered as the center of the RBB.

Finally, we can compute the coordinates of the four corners of the RBB in the image given the width, height, center, and rotation of the box. We proceed by creating an AABB with the determined center, width, and height (width is along the x-axis). Then, we apply a rotation matrix to each of the corner points to get the rotated corner points $\{(x'_i, y'_i), i = 1, 2, 3, 4\}$. For example, if (x, y) is a corner point that is rotated clockwise by an angle θ, we can get the rotated points (x', y') using the equation:

$$\begin{bmatrix} x' \\ y' \end{bmatrix} = \begin{bmatrix} cos\theta & -sin\theta \\ sin\theta & cos\theta \end{bmatrix} \begin{bmatrix} x \\ y \end{bmatrix} \tag{1}$$

Adapting AABBs with Mask R-CNN. We employ the popular Mask R-CNN [15] model pre-trained on the MS COCO 2017 [19] dataset to get the regions of the objects inside the AABBs. Let's assume that we need to determine an RBB proposal for an object of class c and the object is contained in an AABB b. Then, the model is used to run prediction on the whole image and we get segmentation masks as well as respective bounding boxes for objects of class c. Later, we find the predicted AABB that has the maximum overlap with the given bounding box b and utilize the associated segmentation mask to locate each pixel belonging to the encapsulated object in the patch image. Once the region of the object is found, the post-processing steps remain the same as before to get the RBB proposal.

4 Datasets

We chose three datasets of diverse elongated objects (wheat heads, cars, ships) to evaluate our proposed RBB methods. Statistics regarding the annotations in these datasets are summarized in Table 1 and Fig. 3.

Wheat Dataset. The wheat dataset is adopted from the Global Wheat Head Detection (GWHD) dataset [11]. The original dataset is comprised of wheat head images from ten different locations around the world covering a large number of genotypes from Europe, North America, Australia, and Asia. The full dataset consists of 4,698 RGB images with 188,445 wheat heads labeled with AABBs. For our proposal generation experiments, we have randomly sampled images from four of the sources named usask_1 (Canada, 200 images), rres_1 (UK, 200 images), ethz_1 (Switzerland, 200 images), and inrae_1 (France, 176 images). However, the participant study to evaluate manual RBB annotation methods uses images from utokyo_1 and utokyo_2. The selected images are manually annotated with RBBs.

Car Dataset. The car dataset is prepared by selecting drone-view images from the CARPK dataset [16] that contains cars from four different parking lots. The native dataset has 1,448 images with 89,777 cars and only AABB annotations are available. Many of the images are close to identical as the drone used to collect the images was stalled or moving slowly. To reduce the overlap between images, we have taken each 10th image of the dataset and a total of 150 images are selected. Similar to the wheat dataset, selected images are manually labeled with RBBs for further experimentation.

Table 1. Object detection datasets used for RBB studies.

Name	No.of images	No. of labeled objects	Avg no. of objects/image
Wheat	776	30,449	39
Car	150	9,330	62
Ship	4,500	5,401	1

(a) Wheat dataset

(b) Car dataset

(c) Ship dataset

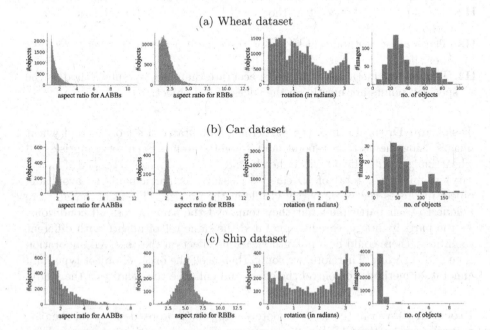

Fig. 3. The distribution of aspect-ratio for AABBs, aspect-ratio for RBBs, rotations of RBBs and number of annotated objects with RBBs for wheat (a), car (b) and ship (c) datasets.

Ship Dataset. The images of the Airbus Ship Detection Challenge dataset [3] are used to form the ship dataset. The challenge dataset has more than 100k satellite images but only around 25% of the images contain ships. First, we process the dataset by discarding the images with no ships. Then, only the images (≈9k) where the area of each of the ships is more than 1,000 pixels are kept. Finally, we randomly select 4,500 images from those to build the ship dataset.

5 Experiments and Results

5.1 Manual RBB Annotation

We conducted a participant study to examine the impact of the proposed RBB drawing methods and the current method on annotation time and accuracy. The study used a 3×2 within-participants RM-ANOVA with factors *Annotation Method* (3-point, 4-point, current) and *Object Type* (car, wheat). The car and wheat instances represent rectangular and ellipsoid shaped objects respectively. The dependent variables were annotation time and accuracy. We analyzed the dependent variables separately. Our hypotheses for the studies were:

H1. The 3-point and 4-point methods will be faster than the current method.

H2. The 3-point and 4-point methods will be more accurate than the current method.

H3. The 3-point method will be faster than the 4-point method as it requires fewer clicks.

H4. The 4-point method will be more accurate than the 3-point method as the specified points are mostly on the edges of the object.

Tasks and Data. Each of the participants annotated a set of car and wheat images using the current, 3-point, and 4-point systems. Each set comprises 3–4 wheat images with 45–60 wheat heads and 3–4 car images (cropped if necessary) containing a total of 50 cars. We asked the participants to avoid any object instances that were partly visible in the images. The set of images were unique to each participant but they remained the same across all conditions. As the participants repeatedly annotated the same set of images with different conditions, there could be a potential learning effect on the observed annotation time and accuracy. Therefore, we counterbalanced the order of object types and annotation methods to control the effect and enhance the validity of the study.

Procedure. We ran the study remotely using the laptops/desktops of the participants due to the COVID-19 restrictions. For each participant, we supplied the images to be annotated and a Windows-compatible installer of the annotation tool. Therefore, we started the study with a device compatibility check. If compatible the participants signed an informed consent form online. Then, the participants filled up a pre-questionnaire which queried them about the level of experience of using computer, mouse, and annotation tool as well as some basic information e.g. gender, age. After completing the form, the annotation tool was installed on their devices. Any required support and guidance were given remotely.

Once the software was installed, the participants were given a demonstration of the user interface of the annotation tool and each of the annotation methods. In addition, the criteria of accurate annotations were explained to them with examples for both wheat and car instances. Then, they were given a few minutes to practice the box drawing approaches on some practice wheat and car images. During the practice tasks, necessary feedback was given to the participants on their annotations over video conferences. They were also instructed to complete the annotations as fast and as accurately as possible for all of the methods.

Finally, the participants performed the main annotation tasks. They took part in two separate sessions. Each session involved annotating either wheat or car images with the three conditions repeatedly. During the sessions, the RBB annotations were saved to measure the accuracy (using IOU) of each box by comparing those to the ground truth RBBs. The annotation time was also recorded for each of the boxes using the annotation tool internally. At the end of every session, the participants completed a user preference questionnaire. They were asked to rank the annotation methods for the recently concluded object type based on the annotation speed (higher rank represented faster method), difficulty (higher rank represented easier method), and overall preference. In total, the study lasted around 90 min for each participant.

Participants and Apparatus. Twelve participants (6 male, 6 female, all right-handed) were recruited from a local university for the user study. The age range of the participants was 24–31 (mean 27.3). The participants self-reported very high familiarity with computer and mouse. However, six of the participants self-reported no experience of using annotation tools, and the others had low to high familiarity (4 low, 1 moderate, 1 high).

The experiments were conducted on the personal laptops of the participants due to the remote nature of the study. All the devices had MS Windows 10 operating system and the display sizes were in the range 15–17 in. The installed annotation tool was written in Python (PyQT5). The annotation inputs were received using a USB optical mouse.

Results and Discussion. For both annotation time and accuracy, we determined outliers within each object type and annotation method for each participant. A trial was considered an outlier and discarded if the measured value was three s.d away from the respective group mean. Out of 1236 trials, 67 trials were discarded based on annotation time and 16 trials were discarded based on annotation accuracy.

ANOVA results showed that for annotation time there was no statistically significant interaction ($F(2, 22) = 2.24, p = 0.130$) between *Annotation Method* and *Object Type*, suggesting that the effect of annotation methods does not depend on the object types. The *Annotation Method* had a significant main effect ($F(2, 22) = 132.09, p = < 0.0001$) on annotation time; Greenhouse-Geiser correction was applied for sphericity violation. Post-hoc pairwise t-tests (Bonferroni corrected) revealed significant difference between the annotation methods (all $p < 0.05$) as shown in Fig. 4. The mean annotation time was 7.54 s (s.d. 2.10 s) for 3-point, 5.78 s (s.d. 1.08 s) for 4-point and 17.3 s (s.d. 3.32 s) for the current method. Therefore, we accept **H1** and must reject **H3**. Both 3-point and 4-point methods were faster than the current method as those required fewer steps. However, the 3-point method involved a higher degree of approximations for ellipsoid shaped objects because distinct object corners are not apparent. Therefore, it performed worse compared to the 4-point method.

RM-ANOVA results indicated no significant interaction between *Annotation Method* and *Object Type* ($F(2, 22) = 4.57, p = 0.15$) for annotation accuracy, again suggesting that the effect of annotation methods does not depend on the object types. We found that there was a significant main effect of *Annotation Method* (($F(2, 22) = 4.44, p = 0.024$) and *Object Type* (($F(1, 12) = 541.39, p =<$ 0.0001) on annotation accuracy. Post-hoc pairwise t-tests (Bonferroni corrected) showed significant difference between the 3-point and 4-point methods ($p = 0.007$) as illustrated in Fig. 4. The mean annotation accuracy was 81.5 (s.d. 5.14) for 3-point, 84.3 (s.d. 5.97) for 4-point and 82.4 (s.d. 5.66) for the current method. Therefore, we must reject **H2** and accept **H4**.

Though the current method was slower, it had a higher accuracy than the 3-point method. It suggests that the participants spent more time making the annotations as accurate as possible. However, we observed during the study that the participants increasingly compromised the accuracy when they had to stay on an object for a longer period. It might be responsible for the lower mean accuracy of the current system compared to the 4-point method but it's not statistically significant ($p = 0.169$).

For annotating car objects, 75% of the participants preferred the 4-point method and the other 25% participants preferred the 3-point method. For wheat instances, all the participants preferred the 4-point method over others. For both types of objects, the current method got the lowest ranking based on the participant preferences.

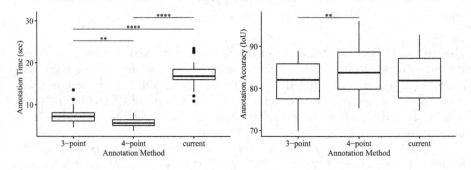

Fig. 4. The effect of annotation method on time (left) and accuracy (right). (** : $p \leq$ 0.01, **** : $p \leq 0.0001$)

5.2 Proposal Generation

We estimate the manual workload to evaluate the different strategies of proposal generation. In the absence of proposal generation, the workload is measured as the number of total objects in the images of a dataset because the annotator will only create a new box for each of the objects. We count the objects of interest by the available ground truth RBBs. When proposal generation is used, the workload is modeled as removals of the incorrect proposals (false positives) and additions of the missed ground truth bounding boxes (false negatives). We calculate the number of such boxes per image using the precision and recall metrics following [6]. When no assisting AABBs is given, the amount of corrections is determined as

$$\# \text{ corrections} = \# \text{ additions} + \# \text{ removals} \qquad (2)$$

where,

$$\# \text{ additions} = \# \text{ objects of interest} \times (1 - \text{recall}) \qquad (3)$$

$$\# \text{ removals} = \# \text{ proposals} \times (1 - \text{precision}) \qquad (4)$$

Also, precision and recall are defined as

$$\text{precision} = \frac{\# \text{ correct proposals}}{\# \text{ generated proposals}} \qquad (5)$$

$$\text{recall} = \frac{\# \text{ correct proposals}}{\# \text{ objects of interest}} \qquad (6)$$

However, if proposals are generated for a set of given AABBs, the number of false positives is limited by the number of given boxes. As the time taken to remove incorrect proposals is minimal in this case, we only examine the number of box additions to determine the number of corrections.

Table 2. Manual workload reduction (%) using different methods on wheat, car and ship dataset. (Higher is better, NA: method does not apply to a dataset)

Detector	Filtering	Adapting	Wheat	Car	Ship
Pretrained	No	No	NA	27.41	64.95
Iterative	No	No	39.77	54.79	69.09
Pretrained	Yes	No	NA	28.75	75.80
Iterative	Yes	No	63.52	66.72	81.04
No	No	Thresholding	50.53	75.11	67.30
No	No	Mask RCNN	NA	4.13	NA
Pretrained	Yes	Thresholding	NA	80.42	87.59
Iterative	Yes	Thresholding	**65.80**	**81.08**	**88.92**

In our experiments, we considered a proposal as correct if the IOU overlap is greater or equal to 0.5 between the ground truth bounding box and the proposed bounding box as well as the difference in the rotations is less than 10°. We configured the other parameters such as confidence score, NMS threshold, and box match IOU threshold to 0.5, 0.1, and 0.2 respectively. The images in a dataset were iteratively traversed to calculate the number of required corrections per image. For iterative detector, proposals were generated on the ith image using the model trained on $1 \ldots (i - 1)$ images for evaluation. Also, the model was trained for 30 epochs for each of the images.

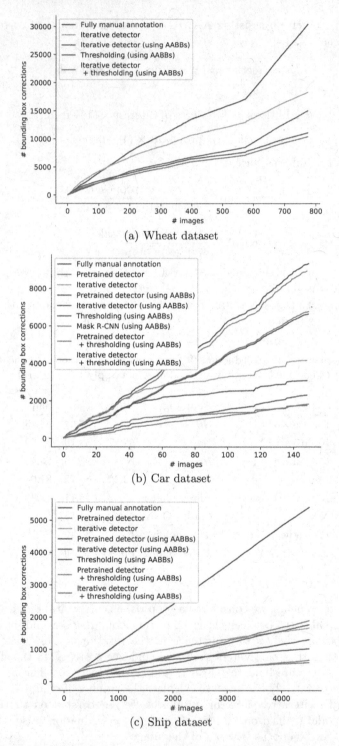

(a) Wheat dataset

(b) Car dataset

(c) Ship dataset

Fig. 5. The required number of manual corrections using different methods for the three datasets. (Lower is better)

The experimental results are shown in Table 2 for all of the datasets. For each dataset, manual workload reduction (%) is calculated as the difference between the cumulative number of ground truth bounding boxes and the cumulative number of required manual corrections for all the images. We see that both the pre-trained and iterative detector get a better performance when proposals are generated for a set of given AABBs. This is because the given boxes can be used to filter out many of the false positives. The maximum workload reduction is achieved for any dataset when the iterative detector is employed in combination with the thresholding technique as the latter can produce proposals for any missed objects by the detector. The Mask R-CNN model experienced a low workload reduction because it is trained on ground-view car images instead of the aerial-view of the car dataset.

Figure 5a reveals the impact of different strategies to alleviate the workload throughout the annotation campaign for the wheat dataset. Without any RBB suggestion, the annotator would need to create a total of 30,420 new bounding boxes. But, with the proposals constructed by the iterative detector in absence of AABBs, the number of manual corrections depletes to 18,321. In presence of AABBs, manual corrections are further reduced to 11,097 bounding boxes. However, it goes to as low as 10,401 giving us a workload reduction of 65.90% when the thresholding approach joins to aid the detector. Figure 5b depicts that the required manual corrections can be as minimum as 1,756 bounding boxes while the total bounding boxes are 9,330. Finally, we can reach the maximum workload reduction of 88.92% as shown in Fig. 5c by using the combination of the iterative detector and thresholding approach. Only 598 boxes need to be corrected out of 5,401 instances in this case.

6   Conclusion

In this work, we extend the publicly available annotation tool, LabelImg, to speed up the tedious task of RBB annotations. We add an online learning based object detector that improves over time to generate RBB proposals. We also integrate a pre-trained detector and segmentation methods to propose RBBs in the early stages of the iterative detector before it is sufficiently trained. The tool provides the user with the flexibility to tune the key parameters of the proposal generation approaches to best match the dataset at hand. We found that the percentage of workload reduction depends on the size and nature of the dataset. Nevertheless, it is evident from our experimental results that our proposal generation methods reduce the annotation time and effort substantially. Furthermore, we propose and evaluate an easy-to-use multipoint drawing system to speed up direct manual RBB annotations. Therefore, we provide an open-source annotation environment that supports efficient and accurate generation of rotated bounding box annotations for large image datasets.

References

1. LabelImg (2016). https://github.com/tzutalin/labelImg. Accessed 18 Sept 2020

2. roLabelImg (2017). https://github.com/cgvict/roLabelImg. Accessed 18 Sept 2020
3. Airbus Ship Detection Challenge (2018). https://www.kaggle.com/c/airbus-ship-detection. Accessed 18 Sept 2020
4. Anno-mage: a semi automatic image annotation tool (2018). https://github.com/virajmavani/semi-auto-image-annotation-tool. Accessed 18 Sept 2020
5. Computer vision annotation tool (CVAT) (2019). https://github.com/openvinotoolkit/cvat. Accessed 18 Sept 2020
6. Adhikari, B., Huttunen, H.: Iterative bounding box annotation for object detection. arXiv preprint arXiv:2007.00961 (2020)
7. Adhikari, B., Peltomaki, J., Puura, J., Huttunen, H.: Faster bounding box annotation for object detection in indoor scenes. In: 2018 7th European Workshop on Visual Information Processing (EUVIP), pp. 1–6. IEEE (2018)
8. Bearman, A., Russakovsky, O., Ferrari, V., Fei-Fei, L.: What's the point: semantic segmentation with point supervision. In: Leibe, B., Matas, J., Sebe, N., Welling, M. (eds.) ECCV 2016. LNCS, vol. 9911, pp. 549–565. Springer, Cham (2016). https://doi.org/10.1007/978-3-319-46478-7_34
9. Biswas, A., Parikh, D.: Simultaneous active learning of classifiers & attributes via relative feedback. In: Proceedings of the IEEE Conference on Computer Vision and Pattern Recognition, pp. 644–651 (2013)
10. Castrejon, L., Kundu, K., Urtasun, R., Fidler, S.: Annotating object instances with a Polygon-RNN. In: Proceedings of the IEEE Conference on Computer Vision and Pattern Recognition, pp. 5230–5238 (2017)
11. David, E., et al.: Global Wheat Head Detection (GWHD) dataset: a large and diverse dataset of high-resolution RGB-labelled images to develop and benchmark wheat head detection methods. Plant Phenomics **2020** (2020). https://doi.org/10.34133/2020/3521852
12. Ding, J., Xue, N., Long, Y., Xia, G.S., Lu, Q.: Learning ROI transformer for detecting oriented objects in aerial images. arXiv preprint arXiv:1812.00155 (2018)
13. Everingham, M., Eslami, S.A., Van Gool, L., Williams, C.K., Winn, J., Zisserman, A.: The pascal visual object classes challenge: a retrospective. Int. J. Comput. Vision **111**(1), 98–136 (2015)
14. Haußmann, M., Hamprecht, F.A., Kandemir, M.: Variational Bayesian multiple instance learning with gaussian processes. In: Proceedings of the IEEE Conference on Computer Vision and Pattern Recognition, pp. 6570–6579 (2017)
15. He, K., Gkioxari, G., Dollár, P., Girshick, R.: Mask R-CNN. In: Proceedings of the IEEE International Conference on Computer Vision, pp. 2961–2969 (2017)
16. Hsieh, M.R., Lin, Y.L., Hsu, W.H.: Drone-based object counting by spatially regularized regional proposal network. In: Proceedings of the IEEE International Conference on Computer Vision, pp. 4145–4153 (2017)
17. Jain, S.D., Grauman, K.: Click carving: segmenting objects in video with point clicks. arXiv preprint arXiv:1607.01115 (2016)
18. Konyushkova, K., Uijlings, J., Lampert, C.H., Ferrari, V.: Learning intelligent dialogs for bounding box annotation. In: Proceedings of the IEEE Conference on Computer Vision and Pattern Recognition, pp. 9175–9184 (2018)
19. Lin, T.-Y., et al.: Microsoft COCO: common objects in context. In: Fleet, D., Pajdla, T., Schiele, B., Tuytelaars, T. (eds.) ECCV 2014. LNCS, vol. 8693, pp. 740–755. Springer, Cham (2014). https://doi.org/10.1007/978-3-319-10602-1_48
20. Liu, W., Ma, L., Chen, H.: Arbitrary-oriented ship detection framework in optical remote-sensing images. IEEE Geosci. Remote Sens. Lett. **15**(6), 937–941 (2018)

21. Papadopoulos, D.P., Clarke, A.D.F., Keller, F., Ferrari, V.: Training object class detectors from eye tracking data. In: Fleet, D., Pajdla, T., Schiele, B., Tuytelaars, T. (eds.) ECCV 2014. LNCS, vol. 8693, pp. 361–376. Springer, Cham (2014). https://doi.org/10.1007/978-3-319-10602-1_24

22. Papadopoulos, D.P., Uijlings, J.R., Keller, F., Ferrari, V.: We don't need no bounding-boxes: training object class detectors using only human verification. In: Proceedings of the IEEE Conference on Computer Vision and Pattern Recognition, pp. 854–863 (2016)

23. Papadopoulos, D.P., Uijlings, J.R., Keller, F., Ferrari, V.: Extreme clicking for efficient object annotation. In: Proceedings of the IEEE International Conference on Computer Vision, pp. 4930–4939 (2017)

24. Parkash, A., Parikh, D.: Attributes for classifier feedback. In: Fitzgibbon, A., Lazebnik, S., Perona, P., Sato, Y., Schmid, C. (eds.) ECCV 2012. LNCS, vol. 7574, pp. 354–368. Springer, Heidelberg (2012). https://doi.org/10.1007/978-3-642-33712-3_26

25. Redmon, J., Farhadi, A.: Yolov3: an incremental improvement. arXiv preprint arXiv:1804.02767 (2018)

26. Ren, S., He, K., Girshick, R., Sun, J.: Faster R-CNN: towards real-time object detection with region proposal networks. In: Advances in Neural Information Processing Systems, pp. 91–99 (2015)

27. Settles, B.: Active learning literature survey. Technical report, University of Wisconsin-Madison Department of Computer Sciences (2009)

28. Tang, T., Zhou, S., Deng, Z., Lei, L., Zou, H.: Arbitrary-oriented vehicle detection in aerial imagery with single convolutional neural networks. Remote Sensing **9**(11), 1170 (2017)

29. Tkachenko, M., Malyuk, M., Shevchenko, N., Holmanyuk, A., Liubimov, N.: Label Studio: data labeling software (2020). https://github.com/heartexlabs/label-studio, open source software available from https://github.com/heartexlabs/label-studio

30. Vondrick, C., Patterson, D., Ramanan, D.: Efficiently scaling up crowdsourced video annotation. Int. J. Comput. Vision **101**(1), 184–204 (2013)

31. Xia, G.S., et al.: DOTA: a large-scale dataset for object detection in aerial images. In: Proceedings of the IEEE Conference on Computer Vision and Pattern Recognition, pp. 3974–3983 (2018)

32. Yang, X., Liu, Q., Yan, J., Li, A., Zhang, Z., Yu, G.: R3det: refined single-stage detector with feature refinement for rotating object. arXiv preprint arXiv:1908.05612 (2019)

33. Zhu, Y., Zhou, Y., Ye, Q., Qiu, Q., Jiao, J.: Soft proposal networks for weakly supervised object localization. In: Proceedings of the IEEE International Conference on Computer Vision, pp. 1841–1850 (2017)

IndianPotatoWeeds: An Image Dataset of Potato Crop to Address Weed Issues in Precision Agriculture

Rajni Goyal[✉], Amar Nath, and Utkarsh

Sant Longowal Institute of Engineering and Technology, Longowal, India
{rajni_pcs2103,amarnath,utkarsh}@sliet.ac.in

Abstract. Over the past ten years, precision agriculture has raised much awareness in the agricultural sector. It automates and optimizes almost all agriculture practices. But the success of this technology depends upon the data. The more accurate and extensive the data is, the more accurate the system will be. Despite large datasets available online, there is still a lack of datasets from the Indian perspective. One of the main roadblocks to advancement is the shortage of publicly available statistics. This paper proposes a benchmark dataset named IndianPotatoWeeds for Potato crops and weeds from Indian farms. The dataset comprises 270 images with annotations and is available online https://www.kaggle.com/datasets/rajni88/indianpotatoweed-dataset. All images were acquired with the Sony CyberShot W830 20.1 M camera and mobile phone. There were intra and inter-row weeds present at the time of data collection. We have provided mask and manual annotation of the plant type (crop vs. weed) for every dataset image using VIA annotation tool. Images can be split into background and foreground via masking, enabling us to concentrate on the areas of the image that interest us. By making this information available to the public, we hope to encourage study in this field.

Keywords: Precision Agriculture · Weed Dataset · Annotation · Image Masking

1 Introduction

Food and fiber are mainly produced via agriculture. Agriculture provides all the world's inhabitants with the nutrition it needs. Also vital to the economy is agriculture. A significant share of the global workforce is employed in the agriculture sector. 50% of the Indian population is employed in the agriculture sector. The economic survey estimates that agriculture's share of the GDP for the fiscal year 2020–2021 was 19.9%[1]. The population has increased exponentially over the past three decades (almost 10 billion by 2050, according to a U.N. study),

---

[1] https://statisticstimes.com/economy/country/india-gdp-sectorwise.php.

© The Author(s), under exclusive license to Springer Nature Switzerland AG 2023
M. K. Saini et al. (Eds.): ICA 2023, CCIS 1866, pp. 116–126, 2023.
https://doi.org/10.1007/978-3-031-43605-5_9

which has led to a sharp rise in food demand [1]. To feed such a vast population, conventional food production methods are insufficient.

Hence, there is a need for innovative and intelligent ways of farming. Smart and Precision farming is one of the best solutions to meet the current food demands of the large population. Precision agriculture provides the automation of various agriculture practices, which reduces time and labor costs. Artificial intelligence(AI) and IoT (Internet of things) are advanced technological tools that help to digitize agricultural activities. To cut waste, boost revenues, and protect the environment, precision agriculture manages each crop production input (water, fertilizer, herbicide, seed, pesticide, etc.) on a site-specific basis. (Ess & Morgan 2013). Precision agriculture has a wide application area. It helps in crop management, weather forecasting, weeds, and pest control, intelligent spraying, livestock farming, remote sensing, storage management, innovative harvesting, etc. It automates almost all agriculture practices. It helps to detect weeds and pests in crop fields and provides site-specific spraying of pesticides and herbicides to remove weeds and pests.

Weeds are a fundamental problem that affects crop yields to a large extent. Weeds are unwanted plants that compete for nutrients, water, and other resources with valuable plants. These unwanted plants are always needed to be removed from fields. The type of weeds depends upon the location, season, and crop. Weeds vary from country to country. So, the kind of weeds present in Germany's land does not need to also be present in the land of India. There are many datasets available from various countries. But from the Indian perspective, there is a lack of weed/crop datasets.

So in this paper, we provide a thoroughly annotated and masked crop/weed dataset from potato fields. The dataset contains 270 images manually annotated using VIA (VGG Image Annotator) [30]. The annotations made available with this dataset enable the development of weed detection and classification solutions and many types of image processing, including edge detection, motion detection, and noise reduction. The information presented is crucial from a computer vision standpoint. On the one hand, the process of picture collecting in the agricultural industry is challenging since it necessitates sophisticated hardware systems, access to fields, and lightning conditions, and the timing of the acquisition must be accurate and linked with the crop growth cycle (only once a year for many cultures). On the other hand, defining appropriate ground truth requires the assistance of agricultural professionals [2].

The dataset comprises field images in a top-down view that were acquired with a Sony CyberShot W830 20.1 M and mobile phone camera. The images are collected from the potato fields of Punjab Agriculture University (Precision farming) fields in Punjab, India. The crop was photographed at a stage of development where many genuine leaves were visible. The manual weeding was done in this field after a few hours of data collection. Here, we focus on potatoes, but wheat, peas, onions, and other cultivars also need manual weed control procedures. Every image has annotations, and the dataset contains crop/weed annotation JSON file, CSV file, Coco format file, and annotation mask for each image.

(a)   (b)

(c)   (d)

(e)   (f)

Fig. 1. Sample images from dataset

The dataset is available online at https://www.kaggle.com/datasets/rajni88/indianpotatoweed-dataset Fig. 1 provides sample images from the dataset.

2   Literature Survey

In general, there is a lack of open datasets accessible by researchers and academicians. Data sets are like the food for classification and detection problems in machine learning models [3]. These technologies are used in a variety of agricultural fields, including crop disease detection, weed classification and identification, plant seedling classification, fruit identification and accounting, management of water resources and soil, weather forecasting (climate) [3–6]. Accurately classifying and detecting weed species in their natural environment may be the most significant barrier to the general adoption of robotic weed management.

The more data included in these databases, the more effective artificial intelligence systems can govern robotic weed growth, provide more accurate plant growth, and allocate scarce resources.Potato/weed dataset [8]is an open-access dataset having 411 images taken from potato fields. But this dataset contains separate images for crop and weed and cannot be used for segmentation problems. It is valid for classification problems only. Another dataset for weed detection has 202 images [9] that can be used for classification problems in deep learning. Another

Table 1. Public Crop/Weed datasets

Dataset	Type of crop	Number of Images	Data Location	Reference
Potato Plant Dataset	Potato	411	not specified	[8]
Weed detection Dataset	not specified	202	not specified	[9]
Crop/Weed Field Image Dataset	Carrot	60	Germany	[10]
Dataset of food crop and weed	Six crop	1118	Latvia	[11]
DeepWeeds	Not Specified	17,509	Australia	[12]
Perennial ryegrass and weed	Perennial ryegrass	33086	USA	[15]
Early crop weed dataset	Tomato, cotton	508	Greece	[14]
Soybean and weed dataset	Soybean	400	Brazil	[16]
Open Plant Phenotype Dataset	Not Specified	7,590	Denmark	[17]
Sugar beet and hegde bindweed dataset	Sugar beet	652	Belgium	[18]
Sugar beet fields dataset	Sugar beet	12340	Germany	[19]
UAV Sugar beet 2015-16 Datasets	Sugarbeets	675	Switzerland	[20]
Corn, Lettuce and weed dataset	Corn and lettuce	6800	China	[22]
Carrot weed dataset	Carrot	39	Republic of Macedonia	[23]
Bccr-segset dataset	Canola, corn, radish	30,000	Australia	[24]
Carrots 2017 dataset	carrots	20	UK	[25]
Onions 2017 dataset	onions	20	UK	[25]
GrassClover image dataset	Red Clover and white clover	31,600 real and 8000 synthetic images	Denmark	[26]
Leaf counting dataset	not specified	9372	Denmark	[27]
CNU weed dataset	not specified	208,477	Republic of Korea	[28]
Plant Seedling dataset	three crops	5539	Denmark	[29]
CNU weed dataset	not specified	208,477	Republic of Korea	[28]
IndianPotatoWeeds	Potato	270	India	This paper

(a)

(b)

Fig. 2. The polygon annotation of images. The yellow color specify crop, and the blue color specify weed.

dataset named cwfid [10], having 60 shots, is available on GitHub for crop /weed classification and segmentation for computer vision in precision agriculture.

Sudras et al. [11] annotated 1118 images having six food crops and eight weed species from different locations in Latvia. DeepWeeds [12] is an extensive dataset having 17,509 images taken from different crop fields in Australia. Table 1 represents the various datasets available online from fields of other countries for different crops.

This paper aims to provide a real-world image dataset for image segmentation and classification model like Faster Region-based Convolutional Neural Network (RCNN) and Mask RCNN. This enables researchers to acquire research on the perception of data acquisition and treatment for weeds in potato fields.

3   Problem Description

Data presented in this paper shows how the dataset is distributed among food crops and weeds. The crop selected for this work is potato. Two hundred seventy images presented in this paper are manually annotated using the VGG image annotator (VIA) tool. The dataset is split into **train** and **val** folders containing 80:20 images. Each folder contains the JSON file having annotations. Raw images and mask for each image is also included in the dataset. Figure 3 displays images from a dataset with polygon annotation with yellow color specifying crop and blue color specifying weed (Better visible in color image).

4   Material and Methods

4.1   VIA (VGG Image Annotator)

VGG Image Annotator (VIA) is an easy-to-use standalone program for manually annotating images, audio files, and videos. There is no setup or installation needed with VIA; it simply runs in a web browser. The complete VIA program is included in a single self-contained HTML page that is less than 400 kilobytes and works as an offline application in most modern web browsers [30]. Using the VIA tool, we have annotated the images. We have also classified images into weed and crop categories, shown in Fig. 3. The region shape used for annotation is the polygon. The total number of annotations is 776, of which 393 are crop annotations and 383 for weed. The extent of the dataset is represented in Table 2.

Table 2. Extent of dataset

Parameter	Value	Format
Image count	270	.jpg
Annotations	776	json, csv
Crop annotations	393	json, csv
Weed annotations	383	.json, .csv
Mask of images	270	.png

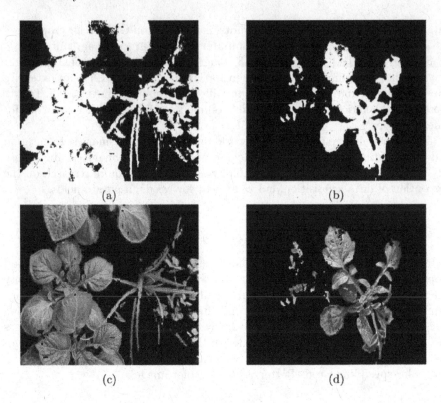

(a)   (b)

(c)   (d)

Fig. 3. Annotated and masked images into crop/weed using VIA tool and python (a) & (b) mask of images (c) & (d) masked images

4.2 Masking

A mask allows us to focus only on the portions of the image that interests us. It can be defined as setting specific pixels of an image to some null value such as 0 (black color). So, only that portion of the image is highlighted where the pixel value is not 0. In this program, we begin with reading the image using the cv2.imread() function in python. Then we convert the image to HSV format as all the operations can only be performed in HSV format.

During masking, the images can be segmented into background and foreground. Figure 3 shows the mask and the masked image from the dataset.

4.3 Field Setup and Acquisition Method

The 270-image dataset was captured at a precision agriculture potato farm in Northern India in December 2022 before manual weeding was applied. The potato plants were grown in a single row on small soil beds. Small close-to-close intra-row weeds were present at data acquisition time. Sony CyberShot W830

Table 3. Specifications of Dataset

Subject	Precision Agriculture, Computer vision, Agronomy, and Science
Specific Subject Area	Image classification, segmentation, Object detection, crop growth and development
Type of data	Images Annotations Image Mask and masked images
Camera specifications	Sony CyberShot W830 20.1 MP and mobile phone camera
Data Format	Raw images: .jpg format Manually annotated images: JSON files Mask and masked image: .png format
Description of data	Dataset consists of Directory Raw images -270 images and mask for each image, train folder - 214 images, JSON file, CSV file test folder - 56 images, JSON file, CSV file
Data source location	Precision farms Punjab Agriculture University, Ludhiana, Punjab, India
Data accessibility	https://www.kaggle.com/datasets/rajni88/indianpotatoweed-dataset

20.1 MP and mobile cameras captured the images in an unregulated environment. During data collection, the weather was clear, with no clouds. Specifications of the dataset are provided in Table 3.

5 Work Flow

Sony Cyber-shot cameras and mobile devices were initially used to capture the raw photos. There were 600 pictures altogether. The data were cleaned to eliminate duplicate photos, blurry images, and noise. After cleaning, 270 images in total were collected. The data were divided 80:20 between train and val folders. Using VIA Annotator, each image was manually annotated. The annotation tool exported JSON and CSV files. We manually constructed a mask for each image and used Python to mask each image. The files were all uploaded to https://www.kaggle.com/datasets/rajni88/indianpotatoweed-dataset.

6 Value of the Data

- The dataset presents images of potato crops and weeds in their early growth stages, which can be used by agronomists and researchers in different fields for computer vision and smart farming.
- The open-access dataset can be used for weed recognition and segmentation algorithms.

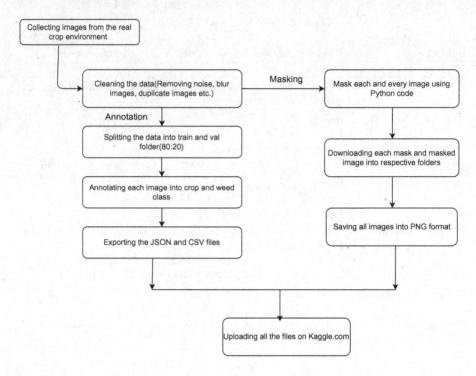

Fig. 4. Work Flow Diagram of the Proposed Approach

– The dataset can train, test and validate convolutional neural networks(CNN) models.

7 Conclusion

A potato crop and weeds dataset for addressing the weed issues in precision agriculture is collected, masked and posted on Kaggle. The images of crops and weeds are acquired using a Sony digital camera and mobile camera in Punjab, India. During the collection of data, there were inter and intra-row weeds were present in the field. The images are manually annotated using VIA (VGG Image Annotator)Tool. There are a total of 270 images in the dataset divided into train and val folders.

This dataset can be used for weed detection, segmentation, and classification problem. We hope this will help increase the progress in the required data acquisition domain and generate ground truth. It will help researchers and agriculture experts to develop ground truth of weed management. In the future, this dataset can be extended with more images from different regions in different seasons and growth days.

Acknowledgement. The authors thank the following colleagues for their comments and help with the acquisition of the dataset: Dr. Rakesh Sharda (Principal Scientist,

Punjab Agriculture University, Ludhiana Punjab (India), Dr. Pankaj Punjab Agriculture University, Ludhiana Punjab (India)).

References

1. Lal, R.: Soil structure and sustainability. J. Sustain. Agric. **1**(4), 67–92 (1991)
2. Haug, S., Ostermann, J.: A crop/weed field image dataset for the evaluation of computer vision based precision agriculture tasks. In: Agapito, L., Bronstein, M.M., Rother, C. (eds.) ECCV 2014. LNCS, vol. 8928, pp. 105–116. Springer, Cham (2015). https://doi.org/10.1007/978-3-319-16220-1_8
3. Yan, J., et al.: Robust multi-resolution pedestrian detection in traffic scenes. In: Proceedings of the IEEE Conference on Computer Vision and Pattern Recognition (2013)
4. Boulent, J., et al.: Convolutional neural networks for the automatic identification of plant diseases. Front. Plant Sci. **10**, 941 (2019)
5. Jeon, W.-S., Rhee, S.-Y.: Plant leaf recognition using a convolution neural network. Int. J. Fuzzy Logic Intell. Syst. **17**(1), 26–34 (2017)
6. Koirala, A., et al.: Deep learning for real-time fruit detection and orchard fruit load estimation: benchmarking of 'MangoYOLO'. Precision Agric. **20**, 1107–1135 (2019)
7. Nkemelu, D.K., Omeiza, D., Lubalo, N.: Deep convolutional neural network for plant seedlings classification. arXiv preprint arXiv:1811.08404 (2018)
8. ALI Hassan Kaggle datasets. https://www.kaggle.com/datasets/ali7432/potato-weed-plants-classification. Accessed 23 Nov 2022
9. AjinJayan. https://github.com/AjinJayan/weed_detection/blob/master/dataset_updated.zip. Accessed 6 Mar 2023
10. Sebastian Haug, Jörn Ostermann github.com. https://github.com/cwfid/dataset. Accessed 23 Nov 2022
11. Sudars, K., Jasko, J., Namatevs, I., Ozola, L., Badaukis, N.: Dataset of annotated food crops and weed images for robotic computer vision control. Data Brief **31**, 105833 (2020)
12. Olsen, A., et al.: DeepWeeds: a multiclass weed species image dataset for deep learning. Sci. Rep. **9**(1), 1–12 (2019)
13. Hasan, A.M., Sohel, F., Diepeveen, D., Laga, H., Jones, M.G.: A survey of deep learning techniques for weed detection from images. Comput. Electron. Agric. **184**, 106067 (2021)
14. Espejo-Garcia, B., Mylonas, N., Athanasakos, L., Fountas, S., Vasilakoglou, I.: Towards weeds identification assistance through transfer learning. Comput. Electron. Agric. **171**, 105306 (2020)
15. Yu, J., Schumann, A.W., Cao, Z., Sharpe, S.M., Boyd, N.S.: Weed detection in perennial ryegrass with deep learning convolutional neural network. Front. Plant Sci. **10**, 1422 (2019)
16. dos Santos Ferreira, A., Freitas, D.M., da Silva, G.G., Pistori, H., Folhes, M.T.: Unsupervised deep learning and semi-automatic data labeling in weed discrimination. Comput. Electron. Agric. **165**, 104963 (2019)
17. Leminen Madsen, S., Mathiassen, S.K., Dyrmann, M., Laursen, M.S., Paz, L.C., Jørgensen, R.N.: Open plant phenotype database of common weeds in Denmark. Remote Sensing **12**(8), 1246 (2020)

18. Gao, J., French, A.P., Pound, M.P., He, Y., Pridmore, T.P., Pieters, J.G.: Deep convolutional neural networks for image-based Convolvulus sepium detection in sugar beet fields. Plant Methods **16**(1), 1–12 (2020)

19. Chebrolu, N., Lottes, P., Schaefer, A., Winterhalter, W., Burgard, W., Stachniss, C.: Agricultural robot dataset for plant classification, localization and mapping on sugar beet fields. Int. J. Robot. Res. **36**(10), 1045–1052 (2017)

20. Chebrolu, N., Läbe, T., Stachniss, C.: Robust long-term registration of UAV images of crop fields for precision agriculture. IEEE Robot. Autom. Lett. **3**(4), 3097–3104 (2018)

21. Madakam, S., Lake, V., Lake, V., Lake, V.: Internet of Things (IoT): a literature review. J. Comput. Commun. **3**(05), 164 (2015)

22. Jiang, H., Zhang, C., Qiao, Y., Zhang, Z., Zhang, W., Song, C.: CNN feature based graph convolutional network for weed and crop recognition in smart farming. Comput. Electron. Agric. **174**, 105450 (2020)

23. Lameski, P., Zdravevski, E., Trajkovik, V., Kulakov, A.: Weed detection dataset with RGB images taken under variable light conditions. In: Trajanov, D., Bakeva, V. (eds.) ICT Innovations 2017. CCIS, vol. 778, pp. 112–119. Springer, Cham (2017). https://doi.org/10.1007/978-3-319-67597-8_11

24. Le, V.N.T., Ahderom, S., Apopei, B., Alameh, K.: A novel method for detecting morphologically similar crops and weeds based on the combination of contour masks and filtered Local Binary Pattern operators. GigaScience **9**(3), giaa017 (2020)

25. Bosilj, P., Aptoula, E., Duckett, T., Cielniak, G.: Transfer learning between crop types for semantic segmentation of crops versus weeds in precision agriculture. J. Field Robot. **37**(1), 7–19 (2020)

26. Skovsen, S., et al.: The GrassClover image dataset for semantic and hierarchical species understanding in agriculture. In: Proceedings of the IEEE/CVF Conference on Computer Vision and Pattern Recognition Workshops (2019)

27. Teimouri, N., Dyrmann, M., Nielsen, P.R., Mathiassen, S.K., Somerville, G.J., Jørgensen, R.N.: Weed growth stage estimator using deep convolutional neural networks. Sensors **18**(5), 1580 (2018)

28. Trong, V.H., Gwang-hyun, Y., Vu, D.T., Jin-young, K.: Late fusion of multimodal deep neural networks for weeds classification. Comput. Electron. Agric. **175**, 105506 (2020)

29. Giselsson, T.M., Jørgensen, R.N., Jensen, P.K., Dyrmann, M., Midtiby, H.S.: A public image database for benchmark of plant seedling classification algorithms. arXiv preprint arXiv:1711.05458 (2017)

30. Dutta, A., Zisserman, A.: The VIA annotation software for images, audio and video. In: Proceedings of the 27th ACM International Conference on Multimedia (2019)

Estimation of Leaf Parameters in Punjab Region Through Multi-spectral Drone Images Using Deep Learning Models

Diksha Arora[✉][iD], Jhilik Bhattacharya[iD], and Chinmaya Panigrahy[iD]

Thapar Institute of Engineering and Technology, Punjab, India
{darora_phd21,jhilik,chinmaya.panigrahy}@thapar.edu

Abstract. This paper reports the use of a deep learning based approach for analyzing leaf micro/macro nutrients from raw multi-spectral drone images. A total of 11 parameters including Boron, Calcium, Copper, Iron, Potassium, Magnesium, Manganese, Sodium, Phosphorus, Sulphur and Zinc are analysed. ResNet-18 model variant has been used for this purpose and the model's precision is assessed using mean absolute error. The study also displays the comparison based analysis of different CNN architectures which includes variants of VGG16, ResNet-50, ResNet-18, AlexNet, LeNet-5, ResNet-152, GoogleNet, and ResNet-101. According to the findings of the study, which was carried out in the fields at the Punjab Agricultural University (PAU), the ResNet-18 model primarily provides better results as compared to other deep learning networks for estimating various leaf parameters.

Keywords: Precision Agriculture · UAVs · ResNet-18

1 Introduction

Agriculture plays a vital role in the economy of every country, particularly in developing nations like India. It is a key area for economic development and the only industry where more than 50% of people work directly in the sector [3]. It does not only provides employment opportunities, but also satisfy mankind's fundamental need for food. By 2050, the world's population is estimated to reach 9.1 billion which represents a significant increase from the current population. As a result, the demand for food is also expected to increase at a rapid pace. Therefore, it is necessary to meet the growing food demands of ever increasing population which can only be achieved through sustainable agriculture [18].

Precision agriculture is a farming method used to achieve sustainable agriculture. It makes use of technology to enhance agricultural operations. Technology can enhance agriculture in multiple ways, both before and after harvesting. One of the ways is by using image processing to analyze soil and leaf nutrient composition, which helps in determining the appropriate quantity and timing

Supported by Thapar Institute of Engineering and Technology, T2CEFS.

for applying farm inputs such as fertilizers, herbicides, water, etc. [1]. Precision agriculture requires gathering and analyzing ample amounts of crop health data, which involve various factors such as water levels, temperature, and other parameters. This involves gathering extensive information from diverse sources and locations within the field, such as soil nutrient levels, crop nutrients, relevant weather conditions, etc. Subsequently, all the collected data is analyzed to generate accurate agronomic recommendations [17].

In recent times unmanned aerial vehicles (UAVs) or drones have proven useful for monitoring and accessing vegetation status. This technology has been found to be particularly effective in monitoring large and remote areas that are difficult to access by foot or ground-based vehicles. With the help of advanced sensors and cameras mounted on the drones, vegetation cover and health can be monitored and analyzed with high precision and accuracy [5]. The availability of UAV-compatible digital RGB, multi-spectral, thermal, and hyper-spectral sensors has greatly improved the ability to conduct in-depth remote sensing investigations in precision agriculture, particularly in the areas of crop yield estimation and field nutrient variability assessment. These sensors, when mounted on UAVs, offer a high degree of control over various parameters such as flight altitude, geographical coverage, acquisition operation, which results in high spatial resolution data [15]. The aim of this study is to analyse the potential of deep neural networks to estimate various crop parameters using multi-spectral drone images.

2   Related Work

Precision farming, which is commonly referred to as digital agriculture, have emerged as new scientific fields that employ data-intensive strategies to increase agricultural productivity [11]. The digital transformation of agriculture has transformed many areas of management into artificially intelligent systems with the purpose of extracting value from an ever-increasing volume of data originating from a variety of sources [4].

Many new and emerging trends in computer science including deep learning (DL) and machine learning (ML) have already been applied in various fields of food and agriculture industry by researchers to solve various complex problems. Machine learning is a rapidly expanding field that focuses on how to create computers that can improve automatically with experience. It combines elements of computer science and statistics, and is a crucial component of data science and artificial intelligence [7]. On the other hand, deep learning is a subset of machine learning that uses neural networks with multiple layers to learn and extract complex features from data. Deep learning is a new and advanced method for image processing and data analysis, which has shown great promise and potential. It has been effectively used in many different areas and has recently been applied in the field of agriculture as well [8]. This advanced technology is helping farmers to minimize losses and gain more by providing them with useful information and insights about crops [14]. For instance, scientific decisions regarding fertilization rely on the foundation of precise and adaptable monitoring of the Nitrogen status of crops.

Table 1. Literature Review

Ref No	Dataset	Location	Pre-process	Objective	Algorithm	GT measure	Crop
[6]	Each experimental plot was scanned to collect hyperspectral data	Northeast of Xiaotangshan Town, Changping District, Beijing	Eliminating noise using Savitzky-Golay (SG) convolution smoothing method	To estimate LNC (Leaf Nitrogen Content)	Partial least squares (PLS) regression, Random forest (RF) and Successive projections algorithm (SPA)	-	Corn
[13]	NDVI maps derived from Satellite images	A vineyard in North Italy	-	Refinement of Vegetation Index driven by Satellite images	RarefyNet	NDVI maps derived from UAV based multi-spectral images	-
[12]	UAV based hyperspectral images	China	Data splicing, radiation correction and geometric correction of hyperspectral orthoimages	Quantitaive estimation of leaf nitrogen content	Back Propagation neural network methods	Spectral data measurements of wheat canopy and leaf sample collection for statistical measurements of LNC	Winter wheat
[16]	UAV based multi-spectral images with five bands of the spectrum	Olive groves located in Portalegre	Radio-metrically corrected, multi-spectral Image Mosaicking and a greyscale DSM generation, background estimation, homogenizes of DSM	To predict leaf parameters i.e. Leaf Phosphorus Content, Leaf Nitrogen Content and Leaf Potassium Content	Partial least squares regression, artificial neural network, Gaussian process regression and support vector regression	Chemical analysis was conducted on leaf samples taken from two different varieties of olive trees	Two varieties of Olive trees

(continued)

Table 1. (*continued*)

Ref No	Dataset	Location	Pre-process	Objective	Algorithm	GT measure	Crop
[19]	UAV based RGB images, multi-spectral images and Raster data construction for crop surface models	The west of Shandong Province in China	Orthomosaic maps were generated for RGB and multi-spectral images	To estimate maize above-ground biomass	DCNN	LAI, AGB, and plant height (PHGM) measurements were done	Maize
[10]	UAV based multi-spectral images with 5 reflectance values	Melbourne, Ontario, Canada	Ortho-mosaic image, Radiometrically corrected	To predict canopy nitrogen weight	Support vector machine, random forests and linear regression using R programming language	For plant tissue analysis, dried biomass weight was sent to A and L canada laboratories and subsequently, LNC was measured	Corn
[20]	UAV multi-spectral imagery which includes five bands i.e., red, green, blue, red-edge, and near-infrared (NIR)	Corn field in southwestern Ontario, Canada	UAV images were orthomosaic using Pix4D-mapper	To predict canopy nitrogen weight	Support vector regression and random forest	Plant height, tissue nitrogen content, dry biomass, soil texture class, soil nitrate nitrogen, water extracted soil nitrate, mineralizable nitrogen, water extracted total nitrogen, and A and L's soil health index rating were evaluated	Corn
[2]	UAV based drone images with four spectral bands, Landsat images having the same UAV fight date were also obtained	The study area located in the Bhakkar district	The SfM (structure from motion) software was used for orthomo-saicking of the drone images	To analyse the Chickpea crop for spatio temporal analysis	Vegetation indices were used to perform a regression analysis on crop growth variables	Measurement of plant height, number of plants and pods, soil pH, temperature and crop yield were recorded in the end of growing season. A field survey was also conducted using a soil-moisture meter and GPS	Chick-pea

Research based on crop nutrients is an important area of study in the field of agriculture and plant sciences. Leaf nutrients are a direct reflection of the nutrient status of the plant as they are the primary site of nutrient uptake and utilization. Measuring leaf nutrient levels can provide important information about the plant's nutritional status, which can be used to diagnose nutrient deficiencies or excesses, and guide fertilizer application and management practices. Additionally, leaf nutrient analysis can be used to monitor the effectiveness of nutrient management practices and assess the impact of environmental factors on plant nutrition. A summarized overview of prior research is shown in the Table 1

In the current study, crop nutrients analysis using multi-spectral images is carried out. The experiments are carried out in the crop intensive yet under explored regions of northern regions of the country. The study is largely motivated by the realization that the outcomes derived from one crop cannot be applied to another due to variations in environmental conditions, soil quality, and crop planting practices. Similarly, the results obtained from one region cannot be reused. The spatial and temporal dependency of the problem, makes it necessary to apply it region and crop wise. The main objective of the study was to explore the potential of deep learning neural networks to predict leaf micro/macro nutrients from multispectral images. The model is trained using pairwise images and leaf analysis data from the lab. As it is already known that leaf analysis for micro/macro nutrients is time taking and costly, hence estimation of the same using multispectral images is very effective for precision agriculture. Automatic leaf analysis using multispectral images will help understand plant nutrition status and can serve as an advisory for fertilizer application. The rest of the paper is organised as follows: Sect. 3 describes the study area, data acquisition and analysis carried out on ground truth data as well as multispectral drone images, Sect. 4 includes the results of the research experiments and Sect. 5 describes the conclusion of the research conducted.

3   Materials and Methods

3.1   Study Area

The experiment was carried out from June 2022 to November 2022 on a paddy field in Punjab Agricultural University (PAU), Ludhiana, Punjab. PR 126 variety of direct seeded rice was sown in the study plot on 9 June 2022. The total area of the experimental plot was 1917 sq. m. Figure 1 shows the experimental field of PAU.

The experimental field was further divided into 10 plots which includes first field as control. Three distinct treatments were carried out in the nine remaining fields, with fields 2, 5, and 8 receiving the same treatment, fields 3,6, and 9 receiving another treatment, and fields 4, 7, and 10 undergoing the third treatment. The variation in treatments were carried out to estimate differences in leaf parameters across the same treatment fields as well as different treatment fields. The reason for applying the same treatment to fields 2, 5, and 8, as well

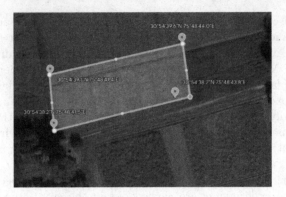

Fig. 1. PAU experimental field

				Field Layout DSR					
Control	Field 1	Field 2	Field 3	Field 4	Field 5	Field 6	Field 7	Field 8	Field 9
3m	6m	6 m	8m	8m	8m	8m	8m	8m	8m

(27m)

Fig. 2. PAU experimental field with field-wise measurements

as to other set of fields was to increase the amount of multispectral image data available for that specific treatment. This strategy was intended to enhance the accuracy of estimating various leaf nutrients. Figure 2 shows the field wise measurements of experimental field in PAU.

3.2 Data Acquisition

Remote sensing using UAVs or drone is expanding globally in a number of agricultural and environmental monitoring and modelling applications. Drones were used to capture multi-spectral images of the PAU field, which consisted of 5 distinct bands including red, green, blue, NIR, and red-edge. The drone was equipped with a Micasense RedEdge-MX multispectral camera whose specifications are given in Table 2. The multispectral camera utilized for data collection has a resolution of 1280*960, whereas the resulting image after orthomosaic has an approximate resolution of 4500*6500. According to the field dimensions, a resolution of approximately 1.4cm/pixel is obtained. In the research analysis, images captured on the following four days were used: [12-Aug-2022, 7-Sept-2022, 20-Sept-2022, and 10-Oct-2022]. The primary objective of the study is to estimate leaf parameters using multispectral image data. To achieve this goal, fertilizer treatments were applied at various stages of crop growth. These fertilizers were absorbed by the soil and subsequently reflected in the crops. Image capturing and leaf sampling were scheduled to take place 2-3 d after each fertilizer treatment. This approach was intended to allow sufficient fertilizer absorption time in the leaves before capturing the multispectral images and collecting leaf samples.

Table 2. Micasense RedEdge-MX multispectral Camera Properties

Pixel Size	Resolution	Aspect Ratio	Sensor Size	Focal Length	Field of View
3.75 μm	1280 × 960	4:3	4.8 mm x 3.6 mm	5.4 mm	47.2° Horizontal, 35.4° Vertical

Leaf samples were also collected from the fields and used as a ground truth for the analysis of different crop parameters. Over the course of four days in the growing season, 10 fields were used for sample collection resulting in 40 groups. Each group is sampled at 8-10 different sampling points. The samples were collected on the same day as the drone acquired multi-spectral images. Figure 3 shows the analysis report of the collected leaf samples. 11 different parameters were analysed from these samples in the testing lab of soil science in PAU. These parameters include Boron, Calcium, Copper, Iron, Potassium, Magnesium, Manganese, Sodium, Phosphorus, Sulphur, and Zinc.

CENTRAL TESTING LABORATORY
DEPARTMENT OF SOIL SCIENCE
PUNJAB AGRICULTURAL UNIVERSITY
LUDHIANA

Test Report (Total Element Composition by ICP-OES)

Element(s)	Sample 1	Sample 2	Sample 3	Sample 4	Sample 5	Sample 6	Sample 7	Sample 8	Sample 9	Sample 10
					mg kg^{-1}					
Boron	2.28	7.63	3.28	5.23	8.03	6.08	7.83	5.83	6.98	8.33
Calcium	2912	2318	2517	2301	3143	3329	2885	2856	2846	2938
Copper	12.40	12.50	11.10	11.70	9.55	12.00	11.90	11.55	8.30	13.50
Iron	81.45	181.35	174.60	217.50	231.50	185.40	461.50	147.35	84.00	306.30
Potassium	20259	19875	19447	18650	20539	18794	19055	17860	17569	18072
Magnesium	1393	1109	1176	1071	1658	1334	1300	1137	987	1223
Manganese	32.03	22.58	25.48	22.38	60.48	20.68	21.83	28.73	19.58	28.33
Sodium	84.15	98.75	90.45	75.40	108	109	103	83.90	100	105
Phosphorous	1811	1664	1662	1619	1937	1894	1954	1717	1570	1609
Sulphur	2497	2294	2355	2379	2774	2956	2788	2913	2670	2842
Zinc	30.88	31.28	26.28	29.63	26.28	29.13	28.43	25.78	21.63	30.93

Fig. 3. Analysis report for collected leaf samples from PAU fields

Data Preprocessing. For each day, a series of multi-spectral images were orthomosaiced, resulting in an orthomosaiced image that is shown in Fig. 4. These orthomosaic images were used to extract two distinct regions of interest (ROIs) from each field. These are generated using Python code. The selection area in each field is random so as to cover different sampling points both in terms of ground truth (leaf sample) collection as well as multispectral data such that the model can be generalized well. As the entire field is covered with the same crop, any area can be treated as a sample region. Each ROI is then further divided into 4 equal parts of 100*100 resolution.

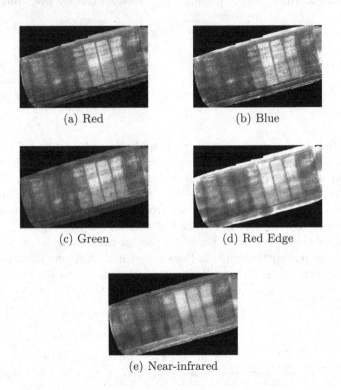

(a) Red   (b) Blue

(c) Green   (d) Red Edge

(e) Near-infrared

Fig. 4. Orthomosaic multispectral images of different bands

3.3 Ground-Truth Data Analysis

In this research the ground truth values of several leaf parameters were compared using a heatmap to examine their relationships. A heatmap is a graphical representation of data that uses colors to display the relative values of each data point in a matrix. Heatmapping enables to easily visualize the patterns and correlations between different leaf parameters. The colors in the heatmap represent the relative values of the parameters. The darker colours denote higher values while lighter colours denote lower values. The heatmap in the Fig. 5 represents the measured correlation matrix, indicating a high correlation between Cu, Fe, K, Mg,

P, and S. The research conducted various experiments to show how this heatmap analysis can improve the understanding of the relationship between various leaf metrics and their corresponding ground truth values. Initially, experiment was conducted on all the leaf parameters, then another experiment was conducted by selecting parameters having low correlation only.

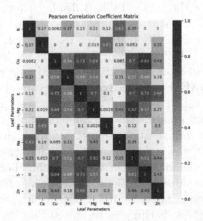

Fig. 5. Heatmap for Leaf Parameters

3.4 Multispectral Drone Data Analysis Using Different CNN Architectures

Convolutional Neural Networks (CNNs) are a form of deep neural networks used most frequently in deep learning to analyse visual data. Modern models for image classification, segmentation, object detection, and many other image processing tasks include convolutional neural networks. To get started in the realm of image processing or to increase prediction accuracy, many architectural techniques have been utilised. Several CNN architectures have been developed over the years, with each one designed to tackle specific challenges. The VGG16, ResNet-50, ResNet-18, AlexNet, LeNet-5, GoogleNet, ResNet-152, and ResNet-101 are among the most popular and widely-used CNN architectures, each with its unique features and capabilities. AlexNet was the first CNN to use the ReLU activation function, while LeNet-5 is designed for handwritten digit recognition tasks. GoogleNet, on the other hand, uses a novel architecture that includes multiple branches to handle different feature scales. VGG16 is known for its simplicity and its ability to extract high-level features from images. ResNet-50 and ResNet-18 use residual blocks to address the problem of vanishing gradients and improve training performance, while ResNet-152 and ResNet-101 are deeper and more complex architectures that achieve state-of-the-art performance on a wide range of computer vision tasks [9].

4 Results

In this research, various CNN architectures have been employed with same hyper-parameters (i.e. epochs=100, learning rate = 0.01) to analyze the prediction of leaf parameters for the purpose of enhancing prediction accuracy. A total of 320 samples were generated from the data using the ROIs. A 70-30 split was used for training and testing the same. Table 3 displays the normalized loss errors observed during the assessment of crop parameters using various deep learning models i.e. VGG16, ResNet-50, ResNet-18, AlexNet, LeNet-5, GoogleNet, ResNet-152, ResNet-101. Different from the general models, the ones used here have 5 input channels instead of 3 and are trained as a regression model with mean absolute error (MAE) loss instead of cross-entropy classification loss. Based on the model's output, either 11 or 6 neurons have been substituted for the last layer. The initial column presents the parameters employed in the analysis. An error rate greater than 1 suggests an unacceptably high error rate. It can be inferred that ResNet-18 yields significantly lower loss than any other architecture when compared to the mean target. This is mainly because ResNet-18 is the shallowest network with skip connections. Other deeper ResNet variants require more data to provide comparative results. Consequently, further research will be conducted using ResNet-18. It can also be stated that an error rate of 0.1–0.12 i.e. 10–12 percent can be termed as acceptable for the current problem. The obtained results show an average of 0.15–0.2 error rates. This model will generalize better with a bigger sample size. It should be noted that the current study was performed with four growth stages collected on four days only.

Table 3. Analysis metrics of Leaf Parameters using different CNN Architectures

Para-meter	VGG 16	ResNet 50	ResNet 18	Alex Net	Le Net5	ResNet 152	Google Net	ResNet 101	Avg Target
B	> 1	> 1	0.57	> 1	> 1	0.55	0.48	> 1	9.32
Ca	0.34	0.13	0.10	0.35	0.17	0.17	0.18	0.17	4478.63
Cu	> 1	> 1	0.29	> 1	> 1	0.57	0.57	0.89	8.16
Fe	0.36	0.28	0.26	0.29	0.28	0.28	0.30	0.29	174.58
K	0.17	0.07	0.05	0.09	0.11	0.09	0.11	0.09	16241.42
Mg	0.25	0.17	0.13	0.19	0.22	0.19	0.20	0.19	1106.77
Mn	> 1	0.24	0.19	0.48	0.98	0.24	0.25	0.33	33.29
Na	0.47	0.25	0.22	0.50	0.42	0.27	0.28	0.29	186.55
P	0.19	0.13	0.09	0.14	0.16	0.14	0.14	0.14	1590.95
S	0.31	0.14	0.13	0.26	0.18	0.17	0.17	0.14	1624.47
Zn	0.82	0.23	0.18	> 1	0.68	0.23	0.31	0.33	25.17

4.1   Estimation of 11 Leaf Parameters Using ResNet-18

In this study, the initial experiment involved the analysis of all 11 parameters using ResNet-18. The 11 parameters were Boron (B), Calcium (Ca), Copper (Cu), Iron (Fe), Potassium (K), Magnesium (Mg), Manganese (Mn), Sodium (Na), Phosphorus (P), Sulphur (S), and Zinc (Zn). The network's input channels were five due to the utilization of multi-spectral images having five distinct bands namely Red, Green, Blue, Near-infrared (NIR), and RedEdge. The model underwent 100 epochs of training and testing, with a learning rate of 0.01. The total number of trainable ResNet-18 parameters were 11,010,699 with 5 input channels and 11 estimation parameters. The FLOPs are approximately 1.82 billion and inference of 25 samples per second are obtained on NVIDIA GeForce RTX 3050 4GB GPU. To assess the accuracy of the predictions, mean absolute error was calculated. The results are depicted in Fig. 6, which displays box plots of the training and testing datasets.

In machine learning, box plots are a crucial tool for data visualisation, especially when used for data analysis and statistics. They are useful for summarizing and displaying the distribution of a data, including the median, quartiles, minimum and maximum values, and outliers. In this research, box plots were employed to visually compare the distribution of various leaf parameters. The study analyzed the normalized loss error ranges of 0–1, encountered by the ResNet-18 model after both training and testing phases. Figure 6 box plot for training phase depicts that Calcium (Ca), Potassium (K), Phosphorus (P), and Sulphur (S) displayed lower loss errors, which indicates better results in comparison to other parameters. It was also observed that Phosphorus (P) and Sulphur (S) exhibited a larger number of outliers. In Fig. 6 box plot for testing phase, it can be seen that Boron (B), Copper (Cu), Iron (Fe), Magnesium (Mg), Sodium (Na) and Sulphur (S) exhibit a lower loss error, while Sulphur (S) shows a higher number of outliers. In addition to this, Fig. 7 shows the MAE loss obtained for both the training and testing phases over 100 epochs while estimating 11 leaf parameters. As the graph shows, both the training and testing losses decrease as the number of epochs increases. This indicates that the model is improving and becoming better at predicting the actual values of the leaf parameters. However, after approximately 20 epochs, the rate of improvement slows down, indicating that the model is approaching its optimal performance level.

Overall, the graph provides valuable insights into the performance of the model in estimating leaf parameters, and the decreasing MAE loss is a positive indication that the model is learning and improving.

4.2   Estimation of 6 Leaf Parameters Using ResNet-18

The model was trained and tested using six parameters: Boron (B), Calcium (Ca), Copper (Cu), Manganese (Mn), Sodium (Na), and Zinc (Zn). These specific parameters were chosen based on a heatmap analysis of all the available parameters, selecting only those with low correlation. Figure 8 illustrates the box plot of loss error during both the testing and training phases. Figure 8 box plot

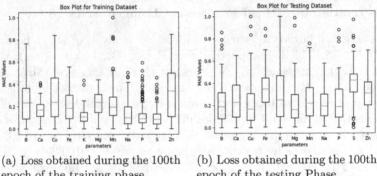

(a) Loss obtained during the 100th epoch of the training phase

(b) Loss obtained during the 100th epoch of the testing Phase

Fig. 6. Loss error with 11 Parameters for both testing and training phase

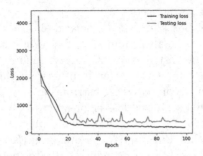

Fig. 7. Loss obtained for 100 epochs while estimating 11 leaf parameters

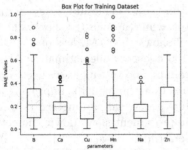
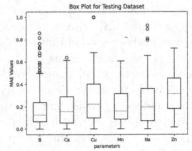

(a) Loss obtained during the 100th epoch of the training phase

(b) Loss obtained during the 100th epoch of the testing Phase

Fig. 8. Loss error with 6 parameters for both testing and training phase

for training phase indicates that Calcium (Ca), Manganese (Mn) and Sodium (Na) displayed lower loss errors in comparison to other parameters. It was also observed that Calcium (Ca), Copper (Cu) and Manganese (Mn) exhibited a larger number of outliers. In Fig. 8, the box plot for the testing phase, it can be seen that Boron (B), Calcium (Ca) and Manganese (Mn) exhibit a lower

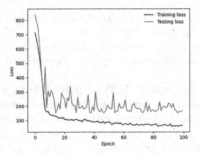

Fig. 9. Loss obtained for 100 epochs while estimating 6 leaf parameters

loss error, while Boron (B) and Sodium (Na) shows a higher number of outliers. Figure 9 shows the MAE obtained for both the training and testing phases over 100 epochs while estimating 6 leaf parameters. Here the graph illustrates a reduction in loss error in less than 10 epochs which indicates great improvement as compared to the above experiment of estimation of 11 different leaf parameters.

5   Conclusion

Based on the research analysis conducted above, it was found that ResNet-18 performs better as compared to other models. The research benefited greatly from the heatmap analysis. Using the data analysis mentioned earlier, experiments were carried out and it was concluded that the model which was trained and tested with only six leaf parameters (i.e. Boron (B), Calcium (Ca), Copper (Cu), Manganese (Mn), Sodium (Na), Zinc (Zn)) showed less error than the model trained and tested with all 11 parameters which are Boron (B), Calcium (Ca), Copper (Cu), Iron (Fe), Potassium (K), Magnesium (Mg), Manganese (Mn), Sodium (Na), Phosphorus (P), Sulphur (S), and Zinc (Zn). The study was conducted on a limited dataset collected between June 2022 and November 2022, which resulted in an average error of 15–20% in estimating leaf parameters. Therefore, to address the limitations of this study, it is necessary to increase the sample size. Currently, the research utilized raw multi-spectral images, but in the future, vegetation indices calculated from these images could also be used for analysis of leaf parameters.

References

1. Abdullahi, H.S., Sheriff, R., Mahieddine, F.: Convolution neural network in precision agriculture for plant image recognition and classification. In: 2017 Seventh International Conference on Innovative Computing Technology (INTECH), vol. 10, pp. 256–272. IEEE (2017)
2. Ahmad, N., Iqbal, J., Shaheen, A., Ghfar, A., Al-Anazy, M., Ouladsmane, M.: Spatio-temporal analysis of chickpea crop in arid environment by comparing high-resolution UAV image and LANDSAT imagery. Int. J. Environ. Sci. Technol. **19**(7), 6595–6610 (2022)

3. Balakrishna, G., Moparthi, N.R.: Study report on Indian agriculture with IoT. Int. J. Electr. Comput. Eng. **10**(3), 2322 (2020)
4. Benos, L., Tagarakis, A.C., Dolias, G., Berruto, R., Kateris, D., Bochtis, D.: Machine learning in agriculture: a comprehensive updated review. Sensors **21**(11), 3758 (2021)
5. de Castro, A.I., Shi, Y., Maja, J.M., Peña, J.M.: UAVs for vegetation monitoring: overview and recent scientific contributions. Remote Sensing **13**(11), 2139 (2021)
6. Fan, L., et al.: Hyperspectral-based estimation of leaf nitrogen content in corn using optimal selection of multiple spectral variables. Sensors **19**(13), 2898 (2019)
7. Jordan, M.I., Mitchell, T.M.: Machine learning: trends, perspectives, and prospects. Science **349**(6245), 255–260 (2015)
8. Kamilaris, A., Prenafeta-Boldú, F.X.: Deep learning in agriculture: a survey. Comput. Electron. Agric. **147**, 70–90 (2018)
9. Krishna, S.T., Kalluri, H.K.: Deep learning and transfer learning approaches for image classification. Int. J. Recent Technol. Eng. (IJRTE) **7**, 427–432 (2019)
10. Lee, H., Wang, J., Leblon, B.: Using linear regression, random forests, and support vector machine with unmanned aerial vehicle multispectral images to predict canopy nitrogen weight in corn. Remote Sensing **12**(13), 2071 (2020)
11. Liakos, K.G., Busato, P., Moshou, D., Pearson, S., Bochtis, D.: Machine learning in agriculture: a review. Sensors **18**(8), 2674 (2018)
12. Liu, H., Zhu, H., Wang, P.: Quantitative modelling for leaf nitrogen content of winter wheat using UAV-based hyperspectral data. Int. J. Remote Sens. **38**(8–10), 2117–2134 (2017)
13. Mazzia, V., Comba, L., Khaliq, A., Chiaberge, M., Gay, P.: UAV and machine learning based refinement of a satellite-driven vegetation index for precision agriculture. Sensors **20**(9), 2530 (2020)
14. Meshram, V., Patil, K., Meshram, V., Hanchate, D., Ramkteke, S.: Machine learning in agriculture domain: a state-of-art survey. Artif. Intell. Life Sci. **1**, 100010 (2021)
15. Misbah, K., Laamrani, A., Khechba, K., Dhiba, D., Chehbouni, A.: Multi-sensors remote sensing applications for assessing, monitoring, and mapping NPK content in soil and crops in African agricultural land. Remote Sensing **14**(1), 81 (2022)
16. Noguera, M., et al.: Nutritional status assessment of olive crops by means of the analysis and modelling of multispectral images taken with UAVs. Biosys. Eng. **211**, 1–18 (2021)
17. Shafi, U., Mumtaz, R., García-Nieto, J., Hassan, S.A., Zaidi, S.A.R., Iqbal, N.: Precision agriculture techniques and practices: from considerations to applications. Sensors **19**(17), 3796 (2019)
18. Sharma, A., Jain, A., Gupta, P., Chowdary, V.: Machine learning applications for precision agriculture: a comprehensive review. IEEE Access **9**, 4843–4873 (2020)
19. Yu, D., et al.: Deep convolutional neural networks for estimating maize aboveground biomass using multi-source UAV images: a comparison with traditional machine learning algorithms. Precision Agric. **24**(1), 92–113 (2023)
20. Yu, J., Wang, J., Leblon, B.: Evaluation of soil properties, topographic metrics, plant height, and unmanned aerial vehicle multispectral imagery using machine learning methods to estimate canopy nitrogen weight in corn. Remote Sensing **13**(16), 3105 (2021)

Application of Near-Infrared (NIR) Hyperspectral Imaging System for Protein Content Prediction in Chickpea Flour

Dhritiman Saha[1]([✉])[ID], T. Senthilkumar[2], Chandra B. Singh[2], and Annamalai Manickavasagan[3]

[1] ICAR-Central Institute of Post-Harvest Engineering and Technology (CIPHET), Ludhiana, Punjab 141004, India
Dhritiman.Saha@icar.gov.in
[2] Centre for Applied Research, Innovation and Entrepreneurship, Lethbridge College, Leth-bridge T1K1L6, AB, Canada
[3] School of Engineering, University of Guelph, Guelph N1G 2W1, ON, Canada

Abstract. Chickpea flour, being high in protein content, is used in several culinary preparations to make protein rich foods. Dumas method is typically used to measure the protein content of chickpea flour, but it is time-consuming, expensive, and labor-intensive. The protein content of chickpea flour was predicted using near-infrared (NIR) hyperspectral imaging in this research. To produce chickpea flour, eight chickpea varieties with varying levels of protein were ground into powder. NIR reflectance hyperspectral imaging was carried out on chickpea flour powder samples between 900 and 2500 nm spectral range. The protein content of twenty-four samples of chickpea flour (8 var × 3 replications) was measured using the Dumas combustion method. The measured reference protein content (dependent variables) and the spectral data (independent variables) of the chickpea flour samples were correlated. Out of total 24 samples, the calibration model was built using 16 powder samples, while the prediction model was built using 8 powder samples. With orthogonal signal correction (OSC)+standard normal variate (SNV) preprocessing, the optimal protein prediction model was obtained using PLSR, which yielded correlation coefficient of prediction (R2p) and root mean square error of prediction (RMSEP) values of 0.934 and 1.006, respectively. Further, competitive adaptive reweighted sampling (CARS) selected 11 feature wavelengths from the studied spectrum and produced the best PLSR model with R2p and RMSEP of 0.944 and 0.889, respectively. As a result, the best prediction model for protein prediction in chickpea flour was obtained by combining PLSR, OSC+SNV and CARS selected wavelength.

Keywords: chickpea flour · near-infrared hyperspectral imaging · regression

© The Author(s), under exclusive license to Springer Nature Switzerland AG 2023
M. K. Saini et al. (Eds.): ICA 2023, CCIS 1866, pp. 141–153, 2023.
https://doi.org/10.1007/978-3-031-43605-5_11

1   Introduction

Pulses are a staple food crop that originates from the leguminous family. The total amount of pulses produced worldwide in 2020 was 89.82 million metric tonnes [1]. Common beans, lentils, chickpeas, dry peas, cowpeas, mung beans, urad beans, and pigeon peas are the principal pulses farmed around the world. Protein content is one of the most important quality variables in pulses [2], and the nutritious value of pulses has boosted demand for its inclusion in baked, milled, and processed foods [3]. The protein content of such food products has a significant effect on their technological efficiency [4]. Chickpea is a legume with a high protein content. After beans and peas, chickpeas are the third most-produced crop in the globe, with a total production of 11.67 million metric tonnes [4]. Chickpea protein content varies from 17% to 24% (dry basis) depending on cultivar, agronomic, and meteorological circumstances [2]. As a result, chickpeas can be a cheap protein source for low-income consumers worldwide, particularly in developing countries where a large population limits access to meat as a protein source. Chickpea protein content is projected to be a component that greatly influences its price in the future, since it will be a critical source for creating protein-enriched goods. Wheat protein content is already employed as a crucial criterion in various nations when determining its price [5].

Although chickpeas can be eaten whole, they often go through many fundamental processing procedures, such as dehulling and grinding to form chickpea flour, which affect the quality, utility, and nutritional content of the proteins in chickpeas [6]. Chickpea flour has a protein content of 17%-21%, a fat content of 5%-7%, a carbohydrate content of 61%-62%, an ash content of 3%, and a water content of 9%-12% [7]. Chickpea flour contains more protein and fiber than wheat flour and is a rich source of polyunsaturated fats. Chickpea flour has been incorporated into various food products, such as bread, pasta, and cakes, along with other cereal flours in recent years, and has been reported to improve the quality of cereal-based products. Chickpea flour is almost always used as a starting point for producing protein-enriched products with appropriate yield, purity, and functional characteristics [8–10]. As a result, chickpea flour production reflects the commercial opportunities that have led to the industrial production of chickpea protein concentrates and isolates from the primary processing of chickpeas into flour. Hence, a rapid and accurate analytical method for detecting the protein content of chickpea flour is required. Wet chemical analysis, specifically the Kjeldahl and Dumas combustion method, is the most used method for assessing chickpea protein content [11]. Conventional chemical treatments, on the other hand, are typically inefficient, pricey, and harmful to the environment.

NIR spectroscopy is combined with digital imaging to provide a three-dimensional "hypercube" dataset with a single spectral dimension and two spatial dimensions for each pixel in the image [12,13]. NIR hyperspectral imaging has a penetration depth of roughly 4 mm, making it suitable for determining interior qualities of food goods [14]. It has been used successfully to determine the chemical composition distribution in a wide range of foods, including fruits, vegetables, meat, fish, and a number of cereal applications [15,16]. Multivariate regression calibration approaches that are effective and highly accurate are crucial in detecting the internal quality of food products and forecasting its composition using HSI. Several chemometric approaches, such as partial least squares regression, support vector machine regression, and artificial neural network, have been developed to identify the composition of chemical components [17]. The most used models for displaying linear and non-linear relationships, respectively, are partial least squares regression (PLSR) and support vector machine regression (SVMR). To reduce modeling complexity and improve model prediction capabilities, feature wavelength selection procedures were used to eliminate collinearity and duplication across HSI data [18]. Furthermore, an adequate and representative detecting position for hyperspectral image acquisition is necessary to minimize positional variability interference in determining the internal content of the component [19].

Even though NIR calibrations work well for detecting protein content in bulk grain samples and are routinely utilized in industry for research laboratories and online measurements, little study on the use of HSI for chickpea protein analysis in flour has been recorded. When paired with the contactless and quick nature of NIR spectrometry, hyperspectral imaging has the potential to increase uniformity in the assessment of chickpea flour and other pulses flour. As a result, the goals of this study were as follows: (1) to investigate the feasibility of using hyperspectral imaging in the Near Infrared (NIR) spectral region (900-2500 nm) to quantitatively predict protein content in chickpea flour; (2) to evaluate the predictive performance of chemometric regression models (PLSR and SVMR) utilizing entire spectra and significant wavelengths acquired using various spectral pre-processing approaches.

2   Materials and Methods

2.1   Chickpea Samples

Eight chickpea varieties harvested in 2020 was supplied by University of Saskatchewan, Saskatoon, Canada. The initial moisture content of all chickpea cultivars was determined using a hot air oven at 105°C for 24 h [11]. The chickpea seeds adjusted to 12.5±0.5% wet basis moisture content was crushed into powder with a Ninja 900W blender and sieved through a Tyler series 50 sieve (300 μm). Using the Dumas combustion method, the protein content of twenty-four (8 var. × 3 replications) samples of chickpea flour was determined for reference. Thereafter, the calibration model was created and validated with the spectral and reference protein data. The imaging and protein content determination of chickpea flour was completed within two days to minimize moisture content fluctuations.

2.2   Hyperspectral Imaging System and Image Acquisition

A camera with spectrograph, a source of illumination (150 W halogen lamp), a translation stage equipped with a sample tray, and a computer running the

Hyperspec III Software suite (Headwall Photonics NIR M series, Massachusetts, USA) comprise the near infrared (NIR) hyperspectral imaging system (Fig. 1).

To obtain the images, 9.50 g of sample was placed in aluminum plates (0.7 cm height and 2.5 cm diameter) and scanned in the 900-2500 nm wavelength range using the reflectance mode. Two plates were scanned at the same time. The thickness of the top layer of the powder top layer was restricted to 7 mm. This thickness meant that the sample holder bottom had no effect on the near infrared reflectance signal of the hyperspectral system [20]. Line by line, the samples were scanned at a speed of 18.01 mm/s. The camera was placed 0.3 m above the samples. The images were captured with the computer's HyperSpec III program (Headwall Photonics). The intensity of the halogen light source was set to 70% during scanning, and the exposure period was set to 7.5 ms. The NIR camera and light source were turned on one hour before image acquisition to ensure thermal and temporal stability. To avoid sample heating, the halogen lamp was only turned on during the scanning phase. The imaging system settings were adjusted to ensure that the samples had the correct aspect ratio and to avoid motor-induced scanning bed vibrations that could perturb the samples. Hyperspectral images were acquired as the scanning bed moved horizontally along the track. The picture hypercubes have 367×368×169 pixels captured at 9.527 nm intervals between 901.121 and 2501.676 nm. Before each imaging session, adjustments were done using black and white references once the images were collected. The camera shutter was closed throughout the dark current measurement, and the white reference was a 99% Spectralon reflectance standard bar (Labsphere, North Sutton, NH) placed below the camera. The images must be corrected for spectral and spatial radiation disparities caused by spectral response, dark current, and unequal light source intensity [21]. The data containing the pixel values were normalized as reflectance using the inbuilt Hyperspec III software (Headwall Photonics, Massachusetts, USA).

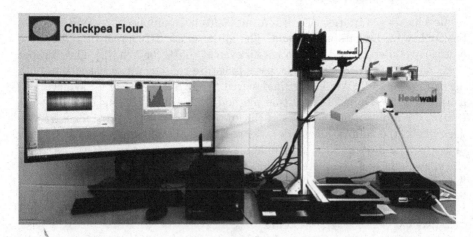

Fig. 1. Hyperspectral imaging system for collecting images of samples

2.3 Quantitative Determination of Protein Content in Chickpea

The protein content of each chickpea flour variety was estimated using Dumas combustion method. According to ISO/TS 16634-2, the protein values were measured on an as-is moisture basis (N× 6.25) (2009). The analysis was performed on a Leco FP-628 (LECO, Stockport, UK) piece of equipment. The samples were analyzed within two days to minimize moisture content change between HSI and Dumas assessments. To test for any drift during the experiment, two samples of chickpea flour with low (16.2%) and high (25.4%) protein contents were used on a consistent basis. The final data were presented in the form of "as-is" total protein content.

2.4 Spectral Data Extraction and Pre-processing

The hyperspectral images of powder samples were in HDR format, which was converted to .mat (MATLAB) files for convenience of processing, and the transformed .mat data were treated to a median filter to remove dead pixels. To label and segment the individual powder samples, the threshold and bwlabel functions were utilized [22]. The powder sample was selected as the region of interest (ROI) after segmenting the corrected hyperspectral images, and a mask was built to discern between the ROI and the background using the difference in intensity. MATLAB 2020a (Mathworks, Natick, USA) code was used to extract the mean spectrum from the pixels within the ROI for each sample [23]. This study's spectral library included twenty-four replicates of pure chickpea flour of eight varieties. Non-useful information, electrical noise, background noise, radio scattering, and baseline drift are common in spectral data. Traditional spectral preprocessing methods, such as Savitzky-Golay (SG) derivatives (1st and 2nd), Standard Normal Variate (SNV), Orthogonal Signal Correction (OSC), Multiplicative Scatter Correction (MSC), and their combinations, were applied to spectral data in order to reduce the effects of these irrelevant elements on modeling and improve model accuracy [24].

2.5 Feature Wavelength Selection

The hyperspectral pictures acquired contained 169 wavelengths, which is a substantial amount and exhibits multi-collinearity, resulting in a longer collecting time. Hyperspectral data with high dimensionality is unsuitable for commercial applications where camera acquisition and data processing speed must match manufacturing pace. To improve calculation speed and prediction accuracy, a subset of critical wavelengths from the entire spectral matrices was chosen to compress hyperspectral data. These wavelengths must offer the majority of the information needed to effectively analyze chickpea flour adulteration. The ROI's spectrum data spans the wavelength range 900-2500 nm, with significant overlap due to the close link between adjacent bands. As a result, it is critical to select the appropriate feature wavelengths while performing spectroscopic analysis in order to construct a robust model with fewer wavelengths [25].

Competitive Adaptive Reweighted Sampling (CARS). To choose effective wavelengths, the Competitive Adaptive Reweighted Sampling (CARS) technique was used. This technique, which was implemented in MATLAB 2020a (MathWorks Inc., Natick, MA), chooses important wavelengths step by step [26]. The absolute value of the regression coefficient from the PLS model is used in CARS to select the N wavelength subgroups using the Monte Carlo approach [27]. The modest weight of the wavelength is then removed using an Exponentially Decreasing Function (EDF) and Adaptive Reweighted Sampling (ARS) approach. This strategy is analogous to Darwin's principle of "survival of the fittest." The feature wavelengths were tested in this study using tenfold cross validation with the number of Monte Carlo sampling runs (N) set at fifty because a higher number of sampling runs did not result in a substantial improvement of the results. Following many loop runs, several subsets of wavelengths were obtained. Finally, effective wavelengths are those with the lowest root mean squared cross validation error (RMSECV) [28]. The chosen wavelengths were used to build the prediction models.

Iteratively Retaining Informative Variables (IRIV). Iteratively Retaining Informative Variables (IRIV) is a variable selection strategy based on the Binary Matrix Shuffling Filter (BMSF) [29]. Model Population Analysis (MPA) is used by the algorithm to classify all variables as highly informative, poorly informative, uninformative, or interfering. Uninformative and interfering variables are eliminated through an iterative procedure. Finally, the feature variables are selected from among the variables that remain following backward elimination [30]. In this investigation, ten-fold cross validation was used with a maximum of ten principal components. Following that, prediction models were built utilizing the wavelengths selected.

2.6 Model Training and Evaluation

PLSR is a linear predictive method for analyzing multivariate data. Many researchers have used it to overcome the difficulty of producing quantitative forecasts in the agricultural domain. The PLSR method combines PCA and multiple regression. In this example, it seeks a collection of latent factors that shed light on the covariance between the two variables. The appropriate number of latent components is determined by the lowest value of the cross-validation root mean square error [28]. In this study, PLSR was utilized to generate linear regression models based on complete spectra and feature wavelengths.

PLSR assumes a linear spectrum-property relationship which is not always true [31], a non-linear model such as Support Vector Machine Regression was used as a comparison. The models were built using ten-section split venetian blind cross-validation. This method of cross-validation divides the dataset into many folds, each with a preset sample count. After determining the number of folds, the classic n-fold cross-validation procedure is used, in which several models are trained with n-1 folds and then tested with the remaining fold.

This technique is repeated until all folds have been used in training except one. The average performances of these models are computed to obtain the venetian cross-validation result. Further, the models' robustness and capacity to predict protein content in samples not contained in the training set were assessed using an independent test set [32].

The Kennard-Stone method was used to randomly partition the original dataset in a 66:34 ratio to create 16 samples for the calibration set and 8 samples for the test set. The Kennard Stone technique [33] is a rigorous sample selection procedure that assures that the samples picked accurately represent the calibration and test datasets. All models used the same calibration set of 16 chickpea flour and prediction set of 8 chickpea flour to avoid bias. The performance of the models was evaluated in this study using the Correlation Coefficient of Calibration (R_c^2), Cross-Validation (R_{cv}^2), and Prediction (R_p^2), as well as the Root Mean Square Error of Calibration (RMSEC), Cross-Validation (RMSECV), and Prediction (RMSEP). To assess the model's reliability, the R2c, R2cv, and R2p were used as guidelines. RMSEC, RMSECV, and RMSEP parameters could reflect the average difference between the predicted and actual values in the relevant set. A good model should have greater R2c and R2p values and lower RMSEC and RMSEP values in general. These parameters were determined using the following formulas:

$$R_c^2, R_{cv}^2, R_p^2 = 1 - \frac{\sum_{i=1}^{n}(y_i - Y_i)^2}{\sum_{i=1}^{n}(y_i - Y_m)^2} \tag{1}$$

$$RMSEC, RMSECV, RMSEP = \sqrt{\frac{1}{n}\sum_{i=1}^{n}(y_i - Y_i)^2} \tag{2}$$

where yi is the measured protein content of the i-th flour in the calibration set and Yi is the predicted protein content of the i-th flour in the prediction set, ym is the mean of all chickpea flour protein content measurements in the calibration or prediction set, and n is the number of flours in the calibration or prediction set. Furthermore, the model's robustness was assessed using Residual Predictive Deviation (RPD), which is the ratio of the Standard Deviation (SD) of the population's reference value to the standard error of prediction in the cross-validated data set. RPD standardizes the prediction accuracy of the model [34], and RPD values of 2.4-3.0, as stated by [35], indicate a bad model; values greater than 3 indicate a robust model.

$$RPD = \frac{SD}{RMSEP} \tag{3}$$

Data extraction was carried out utilizing an in-house MATLAB script. (version 2020a, The Mathworks Inc., Natick, MA, USA). The models were created using the PLS Toolbox (Eigenvector Research, Inc., WA 98,801, USA).

3   Results and Discussion

3.1   Spectral Data Analysis

Figure 2 depicts the original spectral curves of chickpea flour. The spectral curves showed similar tendencies, but their reflectance values differed.

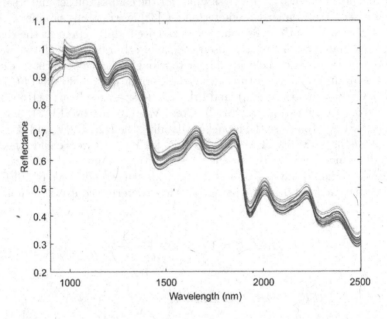

Fig. 2. Average reflectance spectra of chickpea flour with varied protein content

At 1101 nm, the wavelengths corresponded to the second overtone of C-H stretch in carbohydrates. The peak at 1301 corresponds to the N-H stretch's initial overtone. The 1654 nm peak further suggested the C=C aromatic stretch in proteins. The peak at 1854 nm, is the result of a C=O second overtone and a -CONH peptide bond, serving as an indicator for proteins. It is widely known that the C-H in-plane band emerged largely at 1000-1100 nm and 2000-2500 nm (carbohydrates), but the N-H combination band was mostly absorbed around 1500-2000 nm (proteins). The implicit relationship between protein content and spectra was mined and expressed using data analytic methods [11].

3.2   Model Development with Full Spectrum

With pre-processed spectrum data from hyperspectral images of chickpea flour and their corresponding reference protein content, the PLSR and SVMR models covering the complete spectral range were built. The calibration, cross-validation, and prediction findings for protein prediction in chickpea were summarized in

Table 1. R2c, R2cv, and R2p values, as well as RMSEC, RMSECV, and RMSEP values, were used to express model performance. The OSC+SNV was the best pre-processing technique for protein content prediction in chickpea flour using PLSR with four latent variables and an R2p value of 0.934 and an RMSEP value of 1.006. Furthermore, the model built using the OSC+SNV had a high RPD score of 3.718, suggesting the model's robustness. The accurate prediction can be due to the use of spectral pre-processing techniques, including OSC and SNV. OSC tends to remove the anomalies generated from the spectral data whereas SNV seeks to normalize the data. The SVMR also performed well in predicting protein content in chickpea flour with R2p value of 0.886 and an RMSEP value of 1.285.

Table 1. Prediction results of protein content in chickpea flour using full spectrum (900-2500 nm) and PLSR and SVMR with different pre-processing techniques.

Model	Pre-processing	LVs	R_c^2	RMSEC	R_{cv}^2	RMSECV	R_P^2	RMSEP	RPD
PLSR	Nil	4	0.899	1.152	0.860	1.345	0.888	1.123	3.112
	SNV	3	0.912	0.876	0.889	1.056	0.928	1.099	3.568
	OSC	3	0.865	1.129	0.853	1.377	0.882	1.238	2.981
	OSC + SNV	4	0.921	0.913	0.882	1.222	0.934	1.006	3.718
		SVs							
SVMR	Nil	56	0.742	1.731	0.599	2.054	0.573	2.124	1.463
	SNV	45	0.861	1.292	0.723	1.547	0.786	1.619	2.821
	OSC	48	0.842	1.238	0.598	2.217	0.725	1.871	1.889
	OSC + SNV	54	0.925	1.132	0.813	1.123	0.886	1.175	2.911

PLSR - partial least square regression; SVMR- support vector machine regression; SNV - standard normal variate; OSC - orthogonal signal correction; R2c - correlation coefficient of calibration; R2cv - correlation coefficient of cross validation; R2p - correlation coefficient of prediction; RMSEC - root mean square error of calibration; RMSECV - root mean square error of cross validation; RMSEP - root mean square error of prediction; RPD - residual predictive deviation; LV- latent variables. .

3.3 Model Development Using Feature Wavelengths from CARS and IRIV

Feature Wavelengths Selection Using CARS. CARS estimated a total of 11 wavelengths from 169 full-spectrum wavelengths (Table 2). The PLSR model outperformed the SVMR model in terms of prediction ability when the wavelength selected by CARS was used in model development. The calibration, cross-validation, and prediction results for protein prediction in chickpea flour are shown in Table 3. For the optimum PLSR model with OSC+SNV pre-processing, the R2p and RMSEP of the prediction set for protein content were 0.944 and 0.889, respectively.

Table 2. CARS and IRIV selected wavelengths from full spectrum (900 -2500 nm)

Method	Number of wavelengths	Wavelength (nm)
CARS	11	1434, 1444, 1453, 1463, 1472, 1482, 1682, 1691, 1710, 1720, 1996
IRIV·	05	1406, 1853, 1863, 2244, 2263.

Table 3. Prediction results of protein content in chickpea flour using CARS selected wavelengths and PLSR and SVMR with different pre-processing techniques; LV- latent variables; SV-support vectors.

Model	Pre-processing	LVs	R_c^2	RMSEC	R_{cv}^2	RMSECV	R_P^2	RMSEP	RPD
PLSR	Nil	3	0.888	1.155	0.860	1.295	0.908	1.136	3.292
	SNV	3	0.903	1.073	0.880	1.194	0.913	1.116	3.351
	OSC	3	0.886	1.172	0.856	1.323	0.900	1.191	3.140
	OSC + SNV	3	0.925	0.943	0.903	1.073	0.944	0.889	4.207
		SVs							
SVMR	Nil	32	0.699	2.033	0.574	2.325	0.689	2.304	1.623
	SNV	35	0.945	0.815	0.893	1.126	0.946	0.898	4.165
	OSC	37	0.700	2.035	0.630	2.216	0.711	2.249	1.663
	OSC + SNV	38	0.950	0.803	0.863	1.279	0.925	1.048	3.569

Model Using IRIV Feature Wavelengths. IRIV calculated 05 wavelengths from the full spectrum containing 169 wavelengths (Table 2). When wavelengths were chosen using IRIV, PLSR model fared marginally better than the SVMR model.

Table 4. Prediction results of protein content in chickpea flour using IRIV selected wavelengths and PLSR and SVMR with different pre-processing techniques; LV- latent variables; SV-support vectors.

Model	Pre-processing	LVs	R_c^2	RMSEC	R_{cv}^2	RMSECV	R_P^2	RMSEP	RPD
PLSR	Nil	2	0.919	0.985	0.907	1.056	0.910	1.156	3.235
	SNV	3	0.922	0.960	0.912	1.024	0.917	1.083	3.453
	OSC	3	0.915	1.016	0.903	1.084	0.899	1.234	3.031
	OSC + SNV	3	0.910	1.031	0.898	1.103	0.924	1.036	3.610
		SVs							
SVMR	Nil	30	0.546	2.587	0.373	2.852	0.620	2.745	1.362
	SNV	28	0.929	0.925	0.904	1.084	0.923	1.072	3.489
	OSC	27	0.579	2.595	0.509	2.727	0.616	2.796	1.338
	OSC + SNV	31	0.922	0.976	0.867	1.263	0.923	1.075	3.479

The calibration, cross-validation, and prediction results for protein prediction in chickpea flour are shown in Table 4. In the best PLSR model with OSC+SNV pre-processing, the R2p and RMSEP values for the prediction set for protein content were 0.924 and 1.036, respectively. The RPD of the model developed using CARS selected wavelengths was found to be greater than the RPD of the model developed using IRIV selected wavelengths. The reason for better prediction could be attributed to a greater number of selected wavelengths by CARS than IRIV. When utilizing IRIV, some critical wavelengths carrying useful information may be lost unintentionally during the iteration process, which would otherwise be retained during CARS method execution.

Reflectance values are a mixture of multiple molecular vibrations, it is difficult to relate a molecule's vibration to a specific wavelength in NIR spectra [36]. The findings show that these characteristic wavelengths play a major role in predicting protein content in chickpea flour.

4   Conclusion

The goal of this work was to estimate protein content in chickpea flour using near infrared hyperspectral imaging in the spectral region of 900-2500 nm and to compare the performance of the chemometric regression models. Partial Least Square Regression yielded the optimal model with R2p and RMSEP values of 0.934 and 1.006, respectively with OSC+SNV spectral preprocessing. To enable rapid quantification and the system's suitability for commercial use, feature wavelengths were chosen without sacrificing significant spectrum information, demonstrating the current study's novelty. CARS and IRIV were used to select feature wavelengths from the full spectrum. CARS selected wavelengths yielded the optimal model using PLSR with R2p and RMSEP values of 0.944 and 0.889, respectively. Overall, the HSI system can be used for fundamental research goals such as wavelength selection or laboratory scale estimation of quality characteristics, and then a multispectral imaging system equipped with dedicated filters can be built for rapid online prediction. However, the impact of physical and biological product variability in the pulses industry may impede the automation of current technology. Thus, future research will concentrate on the development of multiple databases that take into consideration origin, cultivar, harvest season, and other similar critical characteristics to ensure that the calibration dataset contains appropriate variance. To commercialize the technology, manufacture of multispectral imaging systems with specific filters and implementation of the calibration model may be performed.

References

1. Food and Agriculture Organization (fao). Faostat statistical database of the united nation food and agriculture organization (fao) Statistical division, Rome (2020)
2. Barker, B.: Understanding protein in pulses. Pulse advisor. Saskatchewan pulse growers, p. 1 (2019)

3. Diaz-Contreras, L.M., Erkinbaev, C., Paliwal, J.: Non-destructive and rapid discrimination of hard-to-cook beans. Can. Biosyst. Eng. **60**, 7.1–7.8 (2018)
4. Saha, D., Manickavasagan, A.: Chickpea varietal classification using deep convolutional neural networks with transfer learning. J. Food Process Eng. **45**(3), e13975 (2022)
5. Aporaso, N., Whitworth, M.B., Fisk, I.D.: Protein content prediction in single wheat kernels using hyperspectral imaging. Food Chem. **240**, 32–42 (2018)
6. Grasso, N., Lynch, N.L., Arendt, E.K., O'Mahony, J.A.: Chickpea protein ingredients: a review of composition, functionality, and applications. Compr. Rev. Food Sci. Food Saf. **21**(1), 435–452 (2022)
7. Boye, J.I., et al.: Comparison of the functional properties of pea, chickpea and lentil protein concentrates processed using ultrafiltration and isoelectric precipitation techniques. Food Res. Int. **43**(2), 537–546 (2010)
8. Boye, J., Zare, F., Pletch, A.: Pulse proteins: processing, characterization, functional properties and applications in food and feed. Food Res. Int. **43**(2), 414–431 (2010)
9. Day, L.: Proteins from land plants-potential resources for human nutrition and food security. Trends Food Sci. Technol. **32**(1), 25–42 (2013)
10. Schutyser, M.A.I., Pelgrom, P.J.M., Van der Goot, A.J., Boom, R.M.: Dry fractionation for sustainable production of functional legume protein concentrates. Trends Food Sci. Technol. **45**(2), 327–335 (2015)
11. Sharma, S., Pradhan, R., Manickavasagan, A., Thimmanagari, M., Dutta, A.: Evaluation of nitrogenous pyrolysates by PY-GC/MS for impacts of different proteolytic enzymes on corn distillers solubles. Food Bioprod. Process. **127**, 225–243 (2021)
12. Saha, D., Senthilkumar, T., Singh, C.B., Manickavasagan, A.: Quantitative detection of metanil yellow adulteration in chickpea flour using line-scan near-infrared hyperspectral imaging with partial least square regression and one-dimensional convolutional neural network. J. Food Compos. Anal. **120**, 105290 (2023)
13. Saha, D., Manickavasagan, A.: Machine learning techniques for analysis of hyperspectral images to determine quality of food products: a review. Curr. Res. Food Sci. **4**, 28–44 (2021)
14. Huang, M., Tang, J., Yang, B., Zhu, Q.: Classification of maize seeds of different years based on hyperspectral imaging and model updating. Comput. Electron. Agric. **122**, 139–145 (2016)
15. Gowen, A.A., O'Donnell, C.P., Cullen, P.J., Downey, G., Frias, J.M.: Hyperspectral imaging-an emerging process analytical tool for food quality and safety control. Trends Food Sci. Technol. **18**(12), 590–598 (2007)
16. Saha, D., Senthilkumar, T., Sharma, S., Singh, C.B., Manickavasagan, A.: Application of near-infrared hyperspectral imaging coupled with chemometrics for rapid and non-destructive prediction of protein content in single chickpea seed. J. Food Compo. Anal. **115**, 104938 (2023)
17. Pullanagari, R.R., Li, M.: Uncertainty assessment for firmness and total soluble solids of sweet cherries using hyperspectral imaging and multivariate statistics. J. Food Eng. **289**, 110177 (2021)
18. Sun, J., Ma, B., Dong, J., Zhu, R., Zhang, R., Jiang, W.: Detection of internal qualities of hami melons using hyperspectral imaging technology based on variable selection algorithms. J. Food Process Eng. **40**(3), e12496 (2017)
19. Li, Y., Ma, B., Li, C., Guowei, Yu.: Accurate prediction of soluble solid content in dried hami jujube using swir hyperspectral imaging with comparative analysis of models. Comput. Electron. Agric. **193**, 106655 (2022)

20. Laborde, A., Puig-Castellví, F., Jouan-Rimbaud Bouveresse, D., Eveleigh, L., Cordella, C., Jaillais, B.: Detection of chocolate powder adulteration with peanut using near-infrared hyperspectral imaging and multivariate curve resolution. Food Control **119**, 107454 (2021)
21. Senthilkumar, T., Jayas, D.S., White, N.D.G., Fields, P.G., Grafenhan, T.: Detection of fungal infection and ochratoxin a contamination in stored barley using near-infrared hyperspectral imaging. Biosyst. Eng. **147**, 162–173 (2016)
22. Senthilkumar, T., Jayas, D.S., White, N.D.G.: Detection of different stages of fungal infection in stored canola using near-infrared hyperspectral imaging. J. Stored Prod. Res. **63**, 80–88 (2015)
23. Cruz-Tirado, J.P., Fernández Pierna, J.A., Rogez, H., Fernandes Barbin, D., Baeten, V.: Authentication of cocoa (theobroma cacao) bean hybrids by NIR-hyperspectral imaging and chemometrics. Food Control **118**, 107445 (2020)
24. Florián-Huamán, J., Cruz-Tirado, J.P., Fernandes Barbin, D., Siche, R.: Detection of nutshells in cumin powder using NIR hyperspectral imaging and chemometrics tools. J. Food Compos. Anal. **108**, 104407 (2022)
25. Panda, B.K., et al.: Rancidity and moisture estimation in shelled almond kernels using NIR hyperspectral imaging and chemometric analysis. J. Food Eng. **318**, 110889 (2022)
26. Li, H., Liang, Y., Qingsong, X., Cao, D.: Key wavelengths screening using competitive adaptive reweighted sampling method for multivariate calibration. Anal. Chim. Acta **648**(1), 77–84 (2009)
27. Tao, F., et al.: A rapid and nondestructive method for simultaneous determination of aflatoxigenic fungus and aflatoxin contamination on corn kernels. J. Agric. Food Chem. **67**(18), 5230–5239 (2019)
28. Liu, C., Huang, W., Yang, G., Wang, Q., Li, J., Chen, L.: Determination of starch content in single kernel using near-infrared hyperspectral images from two sides of corn seeds. Infrared Phys. Technol. **110**, 103462 (2020)
29. Yun, Y.-H., et al.: A strategy that iteratively retains informative variables for selecting optimal variable subset in multivariate calibration. Anal. Chim. Acta **807**, 36–43 (2014)
30. Yao, K., et al.: Non-destructive detection of egg qualities based on hyperspectral imaging. J. Food Eng. **325**, 111024 (2022)
31. Balabin, R.M., Lomakina, E.I.: Support vector machine regression (SVR/LS-SVM)-an alternative to neural networks (ANN) for analytical chemistry? comparison of nonlinear methods on near infrared (NIR) spectroscopy data. Analyst **136**(8), 1703–1712 (2011)
32. Kucha, C.T., Liu, L., Ngadi, M., Gariépy, C.: Assessment of intramuscular fat quality in pork using hyperspectral imaging. Food Eng. Rev. **13**, 274–289 (2021)
33. Kennard, R., Stone, L.: Computer aided design of experiments. Technometrics **11**, 137–148 (1969)
34. Mishra, G., Srivastava, S., Panda, B.K., Mishra, H.N.: Rapid assessment of quality change and insect infestation in stored wheat grain using FT-NIR spectroscopy and chemometrics. Food Anal. Meth. **11**, 1189–1198 (2018)
35. Williams, P.C., Sobering, D.C.: How do we do it: a brief summary of the methods we use in developing near infrared calibration. In: Davis, A.M.C., Williams, P. (eds.) Near Infrared Spectroscopy: The Future Waves, NIR Publications, Chichester, pp. 185–188 (1996)
36. Qiao, M., et al.: Determination of hardness for maize kernels based on hyperspectral imaging. Food Chem. **366**, 130559 (2022)

Classification of Crops Based on Band Quality and Redundancy from the Hyperspectral Image

Kinjal Dave[✉][iD] and Yogesh Trivedi[iD]

Institute of Technology, Nirma University, Ahmedabad, India
18ptphde187@nirmauni.ac.in

Abstract. Crop classification from hyperspectral remote sensing images is an effective means to understand the agricultural scenario of the country. Band selection (BS) is a necessary step to reduce the dimensions of the hyperspectral image. We propose a band selection method that takes into account the image quality in terms of a non-reference quality index along with correlation analysis. The optimum bands selected using the proposed method are then fed to the three supervised machine learning classifiers, namely, support vector machine, K-nearest neighbours and random forest. We have also investigated the impact of correlation analysis by showing the comparison of the proposed band selection method with another variant of our method where correlation analysis is not included. The result shows that the crop classification shows better performance in terms of overall accuracy and kappa coefficient when image quality and correlation analysis are both considered while selecting optimum bands. All the experiments have been performed on the three hyperspectral datasets, Indian Pines, Salinas and AVIRIS-NG, which contain major crop classes. The results show that the optimum bands selected using the proposed method provide the highest overall accuracy, equal to 89.63% (Indian Pines), 95.88% (Salinas) and 97.44% (AVIRIS-NG). The overall accuracy shows a rise from +2% to +4% to that of bands without considering correlation analysis. The advantage of this band selection method is that it does not require any prior knowledge about the crop to select the bands.

Keywords: Hyperspectral image · crop classification · band selection · Image quality · correlation

1 Introduction

Crop classification is an inevitable step in planning and managing agriculture worldwide. Definitive crop maps derived from remote sensing image help policymakers to understand many essential things, such as crop yield, growth patterns and crop disease [1] for a larger landscape. So it can be said that remote sensing plays a key role in providing images of the Earth's surface using optical or active

M. K. Saini et al. (Eds.): ICA 2023, CCIS 1866, pp. 154–165, 2023.
https://doi.org/10.1007/978-3-031-43605-5_12

sensors [2–5]. One of the optical remote sensing technology known as hyperspectral imaging has gained its importance in exploiting the different agriculture parameters such as biochemical properties, leaf area index, the moisture level of the crops and biotic stress [6]. Many studies reveal that hyperspectral images can classify crops more accurately than multispectral images [20–23].

Hyperspectral imaging includes more than ten bands per pixel with narrow bandwidth typically ranging from 1 to 15 nm. The most common method of collecting (and representing) hyperspectral imagery is the data cube, with the spatial information being collected in the X-Y plane and the spectral information being displayed in the Z-direction. One way to visualise hyperspectral data is as points on an n-dimensional scatterplot. The information for a certain pixel correlates to its spectral reflectance. Figure 1 represents the reflectance spectra of different crops captured using the Airborne hyperspectral sensor AVIRIS-NG of the Anand district of Gujarat.

Classification of crops from the hyperspectral image is a complex task, as it has numerous bands at different wavelengths, and it is not necessary that every band provides unique information. They may share redundant information, so identifying informative and discriminant bands is a big challenge [7]. Moreover, supervised machine learning classifiers require rich training data for accurate crop classification. Hence, hyperspectral data also deals with Hugh's phenomenon. Hugh's phenomenon occurs when the performance of the machine-learning model decreases for limited training samples. There are two majorly adopted solutions, band selection and feature extraction [8]. The latter approach tends to change the original physics of the hyperspectral data, for example, Principal Component Analysis (PCA). Principal component analysis (PCA) uses eigenvalues to assess the importance of the principle components (PCs) it generates, and data reduction (DR) is performed by choosing PCs with higher eigenvalues. Thus, it loses the original significance of each band. Band selection preserves the significant characteristics of bands by selecting the most informative and non-redundant bands. It can be done using various approaches, such as information-theoretical methods [7,9,10], artificial intelligence models [11] and image quality [12]. Crop classification using the new generation hyperspectral sensors has been presented in [13]. The authors have compared the capability of two new-generation hyperspectral data by classifying seven crops. They also have determined optimum bands to classify the crops using peak and through detection [14]

Despite choosing any one of the approaches, identifying bands with maximum information with minimal redundancy can be considered an open-ended problem. This paper aims to investigate the impact of image quality along with correlation analysis to select optimum bands for crop classification. We have used a Blind/Referenceless Image Spatial Quality Evaluator (BRISQUE) as a spatial quality evaluator that uses a natural scene statistics model [15] of locally normalised luminance coefficients to quantify 'naturalness' using the model's parameters. In order to ensure that only non-redundant bands get selected, we also calculate Pearson's correlation coefficient($corr$) between each adjacent band.

The selected bands are then evaluated on three hyperspectral datasets using different machine learning classifiers Support Vector Machine (SVM), K-Nearest Neighbours (KNN) and Random Forest (RF).

The rest of the paper is organized as follows. Section 2 presents datasets and the proposed Band Selection Method based on Image Quality and Correlation analysis (BSIQCorr). Section 3 presents a discussion of the experimental simulation results. Finally, we conclude our work in Sect. 4.

2 Datasets and Methodology

2.1 Indian Pines Dataset

The AVIRIS (Airborne Visible Infrared Imaging Spectrometer) sensor recorded this data in 1986 in Northwest Indiana, USA [17]. It offers a spatial resolution of 20 m and 224 spectral bands with a spectral resolution of 10 nm, encompassing a spectral range of 400 nm-2500 nm nm. There are sixteen classes which have been shown in the Table 1. Figure 2 shows the Indian pine dataset image in false colour composite along with its ground-truth data.

2.2 Salinas Dataset

This is another hyperspectral benchmark dataset which was captured by the 224-band AVIRIS sensor above the Salinas Valley in California [18]. It is distinguished by its great spatial resolution of 3.7 m. After removing noisy bands, we used a total of 200 bands in the experiments. The sixteen classes include vegetables, barren ground, and vineyard lands Table 1 and Fig. 3.

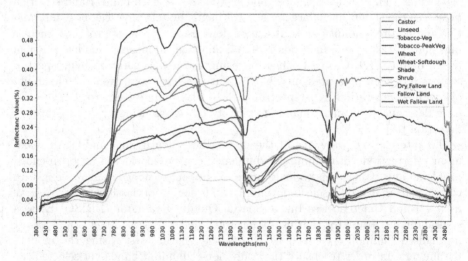

Fig. 1. Reflectance spectral curves of various agricultural classes captured from 380 nm to 2510 nm with 5 nm bandwidth

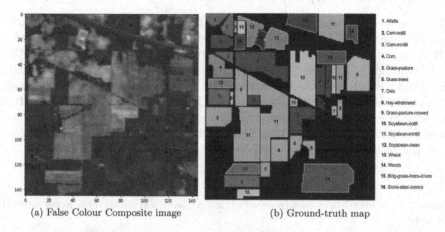

(a) False Colour Composite image

(b) Ground-truth map

Fig. 2. Indian Pines dataset

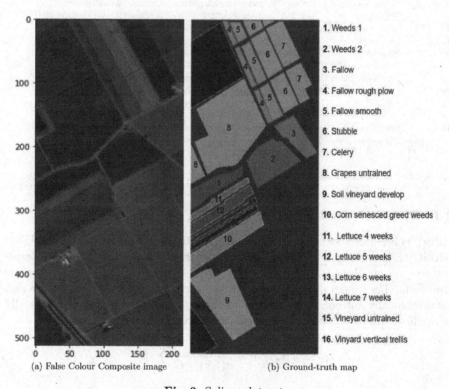

(a) False Colour Composite image

(b) Ground-truth map

Fig. 3. Salinas dataset

2.3   AVIRIS-NG Hyperspectral Dataset

The dataset has been captured using an AVIRIS-NG sensor over the Anand District of Gujarat, India [19]. The area has a heterogeneous agricultural landscape. The Airborne Visible/Infrared Imaging Spectrometer-Next Generation (AVIRIS-NG) campaign is a collaborative project between the Space Application Centre, ISRO and the JPL laboratory of NASA. This airborne sensor provides 425 bands per pixel in the wavelength region of around 380 nm to 2510 nm nm. The spectral and spatial resolution of the sensor is 5 nm and 4 m, respectively. However, we have used 392 bands in our experiments and discarded all noisy bands. A total of eleven agricultural classes have been considered in our experiments.

Table 1. Ground-truth classes (number of training samples) for all three hyperspectral datasets

Class Number	Indian Pines	Salinas	AVIRIS-NG
1	Alfalfa (48)	Brocoli_green_weeds_1 (2009)	Castor (51)
2	Corn_notill (1428)	Brocoli_green_weeds_2 (3726)	Linseed (66)
3	Corn_mintill (830)	Fallow (1976)	Tobacco_vegetative (219)
4	Corn (237)	Fallow_rough_plow (1394)	Tobacco_peak vegetative (107)
5	Grass-pasture (483)	Fallow_smooth (2678)	Wheat (46)
6	Grass-trees (730)	Stubble (3959)	Wheat_softdough (49)
7	Grass-pasture-mowed (28)	Celery (3579)	Shade (127)
8	Hay (478)	Grapes_untrained (11271)	Shrub (103)
9	Oats (20)	Soil_vinyard_develop (6203)	Dry Fellow (295)
10	Soyabean_notill (972)	Corn_senesced_green_weeds (3278)	Fellow (530)
11	Soyabean_mintill (2455)	Lettuce_romaine_4wk (1068)	Wet Fellow (933)
12	Soyabean_clean (593)	Lettuce_romaine_5wk (1927)	-
13	Wheat(205)	Lettuce_romaine_6wk (916)	-
14	Woods (1265)	Lettuce_romaine_7wk (1070)	-
15	Buildings-Grass-Trees-Drives (386)	Vinyard_untrained (7268)	-
16	Stone-Steel-Towers (93)	Vinyard_vertical_trellis (1807)	-

2.4   Proposed Band Selection Method

Blind/referenceless Image Spatial Quality Evaluator (BRISQUE): BRISQUE is an image quality assessment model that uses band pixels to derive features rather than transform them to other spaces, such as discrete cosine transform. The model depends on spatial Natural Scene Statistics (NSS), which contains locally normalised brightness coefficients and their products. The locally normalized luminescence can be calculated using Eq. (1):

$$\hat{I}(i,j) = \frac{I(i,j) - \mu(i,j)}{\sigma(i,j) + C} \tag{1}$$

$$\mu(i,j) = \sum_{k=-K}^{k} \sum_{l=-L}^{L} w_{k,1} I_{k,l}(i,j) \tag{2}$$

$$\sigma(i,j) = \sqrt{\sum_{k=-K}^{k}\sum_{i=-L}^{L} w_{k,l}\left(I_{k,l}(i,j) - \mu(i,j)\right)}, \qquad (3)$$

where, $\omega = \{\omega_{k,l} \mid k = -k \ldots, k, l = -L, \ldots L\}$ is a gaussian kernel of size$(K.L)$, $\mu(i,j)$ and $\sigma(i,j)$ are local mean and deviation respectively. The model uses a generalized Gaussian distribution (GGD) that captures a broader spectrum of distorted image statistics [15]. The model proposes pairwise products of neighbouring MSCN coefficients that can be derived from the Eq. (4).

$$
\begin{aligned}
H(i,j) &= \hat{I}(i,j)\ \hat{I}(i,j+1) \\
V(i,j) &= \hat{I}(i,j)\ \hat{I}(i+1,j) \\
D1(i,j) &= \hat{I}(i,j)\ \hat{I}(i+1,j+1) \\
D2(i,j) &= \hat{I}(i,j)\ \hat{I}(i+1,j-1),
\end{aligned}
\qquad (4)
$$

where $H(i,j)$, $V(i,j)$, $D1(i,j)$ and $D2(i,j)$ are horizontal, vertical, main-diagonal and vertical diagonal orientations as shown in Fig. 4, respectively. As presented in [15], the empirical histograms of products of coefficients do not fit well with the generalized Gaussian distribution. Hence an Asymmetric Generalized Gaussian Distribution (AGGD) model is used. Thus, a total of 18 features for each side has been calculated, as shown in Table 2.

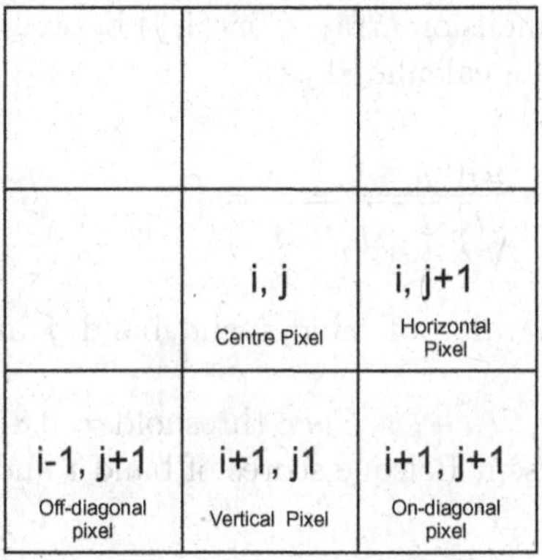

	i, j Centre Pixel	**i, j+1** Horizontal Pixel
i-1, j+1 Off-diagonal pixel	**i+1, j1** Vertical Pixel	**i+1, j+1** On-diagonal pixel

Fig. 4. Quantifying nearby statistical correlations required computing a number of matched products [15]

Table 2. Eighteen features at each side to be fed to a regression model to calculate the Brisque score [15]

Feature ID	Feature Description
$f_1 - f_2$	Shape and variance
$f_3 - f_6$	Shape, mean, left variance, right variance
$f_7 - f_{10}$	Shape, mean, left variance, right variance
$f_{11} - f_{14}$	Shape, mean, left variance, right variance
$f_{15} - f_{18}$	Shape, mean, left variance, right variance

Band Selection Using Image Quality and Correlation Analysis (BSIQ-Corr). First of all, we calculate the score of each band using a non-referenced image quality index BRISQUE. The score can be calculated using the method presented in [15]. It is worth noting that a lower score represents better image quality compared to the band having a higher score. Next, the bands have been sorted by their BRISQUE score in ascending manner. To ensure that the selected bands are discriminant and non-redundant, we calculate Pearson's Correlation Coefficient (r) between each adjacent band pair. We chose to keep the threshold value of r equal to 0.85. Meaning that all band pairs that have the r greater than or equal to the threshold have been considered redundant or correlated bands. It is not necessary to keep both bands in the classification process. Hence, a band with a good BRISQUE score has been chosen as the optimum band. The Algorithm 15 shows the proposed band selection method.

Algorithm 1. BSIQCorr

Input: H is hypercube containing L number of bands with $N \times M$ dimension, where N and M represent number of rows and columns $H = \{b_1, b_2, ..., b_L\}_{N \times M}$
Output: Selected Bandset: Φ

1: **for** i in L **do**
2:      Calculate Brisque score (B_r)
3: **end for**
4: Create an array containing BRISQUE scores of all L bands
5: Sort the array in ascending order based on their BRISQUE scores
6: Calculate Pearson's correlation coefficient($Corr$) between each adjacent band, where each band is flattened to L dimension array. $Corr(i, j)$ between band $i_{(N \times M, L)}$ and band $j_{(N \times M, L)}$ can be calculated as:

$$Corr(i,j) = \frac{\sum_{l=1}^{L}(b_{il} - \bar{b_{il}})(b_{jl} - \bar{b_{jl}})}{\sqrt{\sum_{l=1}^{L}(b_{il} - \bar{b_{il}})^2}\sqrt{\sum_{l=1}^{L}(b_{jl} - \bar{b_{jl}})^2}} \qquad (5)$$

Where b_{il} and b_{jl} represents the spectral value of band i and band j at wavelength l respectively.
7: **if** corr(i,j) $\geq$ 0.85 **then**   ▷ $Corr$ threshold $= 0.85$
8:      **if** $B_{ri} > B_{rj}$ **then** ▷ B_{ri} and B_{rj} represent Brisque scores of band i and j respectively
9:           $\Phi \leftarrow j$
10:     **else if** $B_{rj} > B_{ri}$ **then**
11:          $\Phi \leftarrow i$
12:     **end if**
13: **else**
14:     $\Phi \leftarrow (i, j)$
15: **end if**

---

3   Results and Discussion

We apply the proposed algorithm on three hyperspectral datasets using the three supervised classifiers, SVM, KNN and RF. We have shown the results with another variant of the proposed method, named BSIQ (Band selection using only Image Quality), where step 6 and step 7 were skipped and thus did not include correlation analysis in the band selection process.

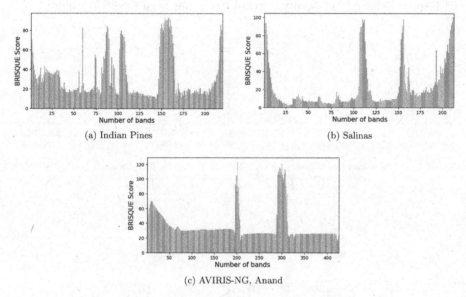

(a) Indian Pines   (b) Salinas

(c) AVIRIS-NG, Anand

Fig. 5. BRISQUE score of each band for the hyperspectral datasets

As the first step of the proposed method is to calculate the BRISQUE score of each band. For example, Fig. 5a represents the score for the Indian Pines dataset. Here, X-axis shows the band number ranging from 400 nm to 2500 nm nm while the y-axis represents the obtained BRISQUE score. The plot has been shown for all 220 bands and hence also includes noisy bands. It can be seen that band numbers 104 to 108, 150 to 163 and 219 have higher BRISQUE scores compared to other bands. Similarly, Fig. 5b and 5c depicts BRISQUE scores of the full band salinas and AVIRIS-NG dataset, respectively. The next step is to calculate *Corr* between each adjacent band. We have selected twenty optimum bands which are informative(i.e. good image quality) and non-redundant (least correlated).

Optimum bands selected using the proposed method (BSIQCorr) have been shown in the Fig. 6a. It is worth noting that the selected bands cover visible to near-infrared regions of the spectrum. The next step is to feed the selected bands in the machine learning classifiers. For SVM classifier, the Radial Basis Function (RBF) kernel was used as it is well known that they are particularly efficient for pixel-based categorization of remotely sensed data [16]. The choice of model parameters has a significant impact on an SVM model's accuracy. The cross-validation process was used to identify the best value of C and gamma for each data set. For KNN, the number of nearest neighbours was kept equal to 3, and in the RF classifier, the number of trees and the maximum number of features in each node was selected equal to 100 and the square root of the total number of input features, respectively. Each experiment has been evaluated using 5-fold cross-validation. A straightforward comparison using the Overall Accuracy (OA) and Kappa coefficient on twenty selected bands has been shown in Figure 7.

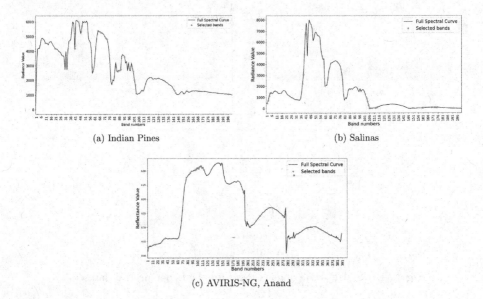

(a) Indian Pines

(b) Salinas

(c) AVIRIS-NG, Anand

Fig. 6. Optimum twenty bands selected using proposed algorithm

The following observations can be drawn from the classification results shown in Fig. 7.

- Indian Pine dataset: Highest performance is obtained by the BSIQCorr using SVM classifier with OA and Kappa coefficient equal to 89.63% and 0.87, respectively. Overall the BSIQCorr is outperforming the BSIQ in both evaluation parameters.
- Salinas dataset: Similar trend can be seen where the BSIQCorr is providing at least +1.5% than that of BSIQ. The highest accuracy and kappa were observed using SVM classification.
- AVIRIS-NG dataset: Optimum twenty bands able to classify the eleven classes with the highest accuracy equals 97.22% and kappa coefficient 0.95 using SVM classifier.

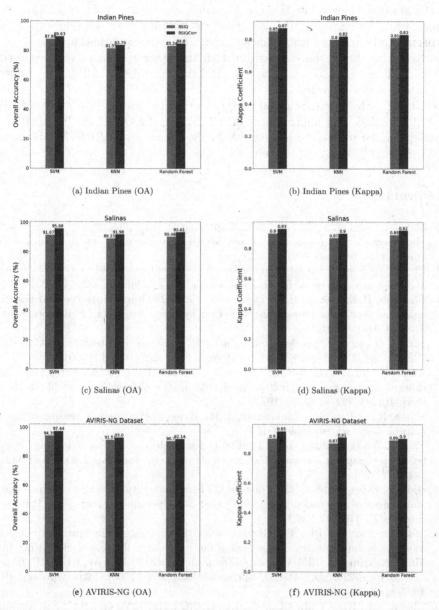

Fig. 7. Classification result using twenty optimum bands selected using the proposed algorithm (BSIQCorr) showing comparison with the method that only considers Image quality (BSIQ)

4   Conclusion

In this study, we looked into how image quality affected the choice of informative bands while taking correlation analysis into account. The outcomes of three hyperspectral datasets with significant agricultural classifications are displayed. We have also shown the comparison with the other variant of the proposed method where correlation analysis is not being considered. All of the results indicate that the suggested band selection strategy delivers the best outcome in accordance with the experimental design. For instance, using BSIQCorr instead of BSIQ results in an improvement of +1.75 % in the OA of the Indian Pines dataset. Similar to this, the highest OA for the Salinas and AVIRIS-NG datasets is 95.88% and 97.44%, respectively.

References

1. Zhu, L., Radeloff, V.C., Ives, A.R.: Improving the mapping of crop types in the Midwestern US by fusing landsat and MODIS satellite data. Int. J. Appl. Earth Obs. Geoinf. **58**, 1–11 (2017)
2. Chen, Y., et al.: Mapping croplands, cropping patterns, and crop types using MODIS time-series data. Int. J. Appl. Earth Obs. Geoinf. **69**, 133–147 (2018)
3. Chauhan, H.J., Mohan, B.K.: Development of agricultural crops spectral library and classification of crops using hyperion hyperspectral data. J. Remote Sens. Technol. **1**(1), 9 (2013)
4. Skriver, H., et al.: Crop classification using short-revisit multitemporal SAR data. IEEE J. Sel. Top. Appl. Earth Obs. Remote Sens. **4**(2), 423–431 (2011)
5. Khosravi, I., Alavipanah, S.K.: A random forest-based framework for crop mapping using temporal, spectral, textural and polarimetric observations. Int. J. Remote Sens. **40**(18), 7221–7251 (2019)
6. Sahoo, R.N., Ray, S.S., Manjunath, K.R.: Hyperspectral remote sensing of agriculture. Curr. sci. 848–859 (2015)
7. Yan, Y., Yu, W., Zhang, L.: A method of band selection of remote sensing image based on clustering and intra-class index. Multimedia Tools Appl. **81**(16), 22111–22128 (2022)
8. Hidalgo, D.R., Cortés, B.B., Bravo, E.C.: Dimensionality reduction of hyperspectral images of vegetation and crops based on self-organized maps. Inf. Process. Agric. **8**(2), 310–327 (2021)
9. Martínez-Usó, A., Pla, F., García-Sevilla, P., Sotoca, J.M.: Automatic band selection in multispectral images using mutual information-based clustering. In: Martínez-Trinidad, J.F., Carrasco Ochoa, J.A., Kittler, J. (eds.) CIARP 2006. LNCS, vol. 4225, pp. 644–654. Springer, Heidelberg (2006). https://doi.org/10.1007/11892755_67
10. Chang, C.I., Kuo, Y.M., Chen, S., Liang, C.C., Ma, K.Y., Hu, P.F.: Self-mutual information-based band selection for hyperspectral image classification. IEEE Trans. Geosci. Remote Sens. **59**(7), 5979–5997 (2020)
11. Sawant, S.S., Manoharan, P., Loganathan, A.: Band selection strategies for hyperspectral image classification based on machine learning and artificial intelligent techniques-Survey. Arab. J. Geosci. **14**, 1–10 (2021)

12. Sun, K., Geng, X., Ji, L., Lu, Y.: A new band selection method for hyperspectral image based on data quality. IEEE J. Sel. Top. Appl. Earth Obs. Remote Sens. **7**(6), 2697–2703 (2014)
13. Aneece, I., Thenkabail, P.S.: New generation hyperspectral sensors DESIS and PRISMA provide improved agricultural crop classifications. Photogram. Eng. Remote Sens. **88**(11), 715–729 (2022)
14. Kutser, T., et al.: Remote sensing of black lakes and using 810 nm reflectance peak for retrieving water quality parameters of optically complex waters. Remote Sens. **8**(6), 497 (2016)
15. Mittal, A., Moorthy, A.K., Bovik, A.C.: No-reference image quality assessment in the spatial domain. IEEE Trans. Image Process. **21**(12), 4695–4708 (2012)
16. Fauvel, M., Benediktsson, J.A., Chanussot, J., Sveinsson, J.R.: Spectral and spatial classification of hyperspectral data using SVMs and morphological profiles. IEEE Trans. Geosci. Remote Sens. **46**(11), 3804–3814 (2008)
17. Gualtieri, J.A., Cromp, R.F.: Support vector machines for hyperspectral remote sensing classification. In: 27th AIPR workshop: Advances in Computer-Assisted Recognition, vol. 3584, pp. 221–232. SPIE (1999)
18. Gualtieri, J.A., Chettri, S.R., Cromp, R.F., Johnson, L.F.: Support vector machine classifiers as applied to AVIRIS data. In Proceedings of the Eighth JPL Airborne Geoscience Workshop, pp. 8–11 (1999)
19. Nigam, R., et al.: Crop type discrimination and health assessment using hyperspectral imaging. Curr. Sci. **116**(7), 1108–1123 (2019)
20. Mariotto, I., Thenkabail, P.S., Huete, A., Slonecker, E.T., Platonov, A.: Hyperspectral versus multispectral crop-productivity modeling and type discrimination for the HyspIRI mission. Remote Sens. Environ. **139**, 291–305 (2013)
21. Vali, A., Comai, S., Matteucci, M.: Deep learning for land use and land cover classification based on hyperspectral and multispectral earth observation data: a review. Remote Sens. **12**(15), 2495 (2020)
22. Wan, S., Wang, Y.P.: The comparison of density-based clustering approach among different machine learning models on paddy rice image classification of multispectral and hyperspectral image data. Agriculture **10**(10), 465 (2020)
23. Adam, E., Mutanga, O., Rugege, D.: Multispectral and hyperspectral remote sensing for identification and mapping of wetland vegetation: a review. Wetlands Ecol. Manage. **18**, 281–296 (2010)

Automated Agriculture News Collection, Analysis, and Recommendation

Shaikat Das Joy[✉] and Neeraj Goel

Department of Computer Science and Engineering, Indian Institute of Technology Ropar, Rupnagar, Punjab, India
{2021csm1008,neeraj}@iitrpr.ac.in

Abstract. A country like India mainly depends on the sector of agriculture. Most people's economies are intensely engaged in the field of agriculture. So, developing the agriculture sector will be an excellent benefit for any country. Nowadays, People can immediately find any solution regarding agriculture through technology's modernization. We can get any news from online articles anytime without any movement. Agriculture news should also be available in online news articles so that people who are intensely engaged with the agriculture field and economy can quickly get their valuable news. People must go through many online news sites to gather all the agriculture-related news. We have proposed an NLP-based solution so people can get all agriculture-related news in one place combining multiple features. In this process, we have collected many articles from multiple online newspapers and classified the agriculture news articles. For the classification process, we have applied several classification models. We have also added a machine learning-based model to check the duplication between news articles. Although, there will be multiple categories of agriculture news so that people can directly follow the news as they want. People will also be recommended articles based on content and times. So, Getting information about agriculture will be more straightforward for the farmer, and they can know about new technologies to apply in their work. Finally, in this proposed work, people can get all the essential agriculture news from various sources in one central point, including many exciting features.

Keywords: NLP · news articles · BERT · classification · Recommendation

1 Introduction

Agriculture news is vital to any country's development or economic system. It is essential for those whose primary income source is agriculture. Not only those people but also many people all over the country follow the updated agriculture news. Even the government is also started many agriculture-based projects to keep developing in this sector. People whose workspace is entirely in agriculture need to know all the recent updates about crops, the environment, market prices, new technologies, government schemes, subsidy news, etc. To get all this,

M. K. Saini et al. (Eds.): ICA 2023, CCIS 1866, pp. 166–179, 2023.
https://doi.org/10.1007/978-3-031-43605-5_13

they must go through many online news sites daily, which takes a long time to find the news they want. Not only countries' news but also the situation and technology of international cultures need to know. However, we all know that a significantly less amount of agriculture news is available in the newspaper. Some online newspapers miss important agriculture news because they mainly emphasize the other news to make people more interested. There are many newspapers based on agriculture news. Those papers have many sections of agriculture news. But all the processes have been done manually, not an automated process. Nowadays, machine learning methods create a massive difference in the updation of the world. In this modern era, our proposed idea will help to get all agriculture news in one place.

People must face some challenges when they want to gather all agriculture news in one place. Not all newspaper shows all information together. As a result, people need to visit several newspapers to get all the agriculture-related news together. In other newspapers, working processes are not automated for agriculture news, so many human resources are needed to fulfill the process. In many online newspapers, we can see old news kept for many days. As a result, the same articles are repeated multiple times. Facing duplicate articles several times is another challenging issue for people.

In this work, we are trying to give a solution to all these problems. We aim to provide a platform for all agriculture news, including many features. Thus people can be helpful by this they can get whatever news related to agriculture in one place. In our approach, there will be different sectors to go to that category directly, and people can search as they want. As a result, the requirement of knowing the information will be more straightforward. We have collected much news on agriculture and non-agriculture from popular news sites. In this work, we classify the news and have created some subcategories of agriculture news. We also monitor that the same news should not be repeated, so we check similarities among news articles. Here we will recommend people's news based on time sensitivity, as an example of which news can show after its dates expire and which news to show on that limited days. Our work analyses the sentiment of the news based on the content. We have used many machine learning-based models in this work, such as SVM, Decision tree, Naive Bayes, Random Forest, and a state-of-the-art model BERT transformer. The overall aim is to gather all agriculture news in one place in a systematic way. This work makes the following contributions:

- Based on Crawler, we have fetched many news articles from the online news site.
- Classify the news between agriculture and non-agriculture.
- Similarity check between articles to remove duplicate articles.
- Sub-classification to make separate categories among agriculture news.
- Recommend news based on content like sentiment analysis and time-sensitive

The remaining parts of the work are organized as follows: Firstly, in Sect. 2, we discuss the related works. Then, in Sect. 3, we describe the methodology of this proposed work and discuss about the dataset. We have shown the result in Sect. 4. Finally, in Sect. 5, the last section contains the conclusion, including future work.

2   Related Works

There are a lot of related works related to our proposed approach. Recent surveys show that researchers will mostly make things automated by using modern technologies. If we come to the classification system. Text classification has been used in many works, and many good classifiers can detect well in recognizing data. SVM (Support Vector Machine) is used in many works and achieved better results than many algorithms in classification. In research, text classification is used for news, and report filtering in medical issues [1]. They found that SVM works significantly better than other algorithms, such as Naive, CNN, KNN, and decision trees. SVM classification works better for two dimensions events. In Facial express detection, SVM works well in research work [2]. Zhang et al. proposed an idea for classifying news articles into several categories [3]. Furthermore, they classified news among agriculture, sports, crime, business, and many terms. In this work, SVM was used as a classification model.

In a crime-based work, authors classified the location after collecting the data from an online news site by using a decision tree classifier in this work. Jijo et al. [4] classified text, smartphones, diseases, and media. In this work, authors have discussed different techniques of this decision tree, such as CART, QUEST, and many more [4]. BERT is another pre-trained model based on Transformer and is considered a state-of-the-art model with the best output result compared to most classification algorithms and works better for a large data set. From various class data, it can detect every class more accurately. In one survey report, Carvajal et al. Illustrated the BERT accuracy compared to other traditional approaches. It outperforms many classification models [5]. Zhang et al. have said BERT can work perfectly if the dataset is well enough to label; thus, the model can identify [6]. In his case, he took the dataset from COCO and labeled it in the best way. Including classification, it works better in some work models, such as summarizing an article by understanding it and finding the similarity in the answer. For document classification, in some cases, BERT has shown a beneficial impact [7]. BERT has two classes, BERT large and BERT base. In a work, authors have tried both of these for their working model and tried to improve the model with their idea for the classification between documents. Qasim et al. proposed work for fake news or article detection [8]. During COVID-19 time, much fake news spread all over the world. Moreover, this news creates a hamper for the public. The authors took the dataset and pre-processed this, then used TF-IDF for data conversion and finally went for the classification approaches. So, the authors tried nine transfer learning models to classify fake news or tweets from social media. Some of those models are BERT large, BERT base, Albert, Roberta, and DistilBERT, among the highest performance shown by BERT large and BERT base. In another work, to identify the Chinese medicines, the BERT model was used to represent text and to classify the medical events, TEXT-CNN was used in this case [9]. BERT and SVM are also used together as a hybrid model to classify the sentiment analysis to get the best performance with accuracy [10]. Munikar et al. have used the pretend BERT model on the SST dataset to classify people's sentiments [11]. In this work, authors have introduced a new model named BAE [12]. Here authors checked the written paragraph's coherence and grammatical mistake. Garg and Ramakrishnan's work proved that the BAE model is more robust than the general NLP-based model. Also, in some Korean projects [13], BERT is used as a text classifier for technical documents.

We have surveyed some work, and some functions are similar to our proposed work. In a work, Gupta et al. have an end-to-end model to find where most crime works are happening [14]. In this work, they have classified the news between crime and non-crime by using two algorithms. In one algorithm, they have classified by taking the article text only; in the other, they took both the article text and title. They have done duplication checks between news using TF-IDF and finally made a framework to avoid the most crime-related places. In a work, they also extracted the location of Srilanka based on the crime articles, and for classification, they used SVM and used Simhash method for similarity checking [15]. In another work, authors classified news and used the Decision tree method for classification [16]. Here in Table 1 we can see some features comparisons between related works.

Table 1. Comparison with the related works

Work	Focus	Classification Model	Recommendation	Time Sensitive Analysis	Duplication Detection
[10]	Sentiment Analysis	BERT+SVM	No	No	No
[13]	Technical Doc. Classification	BERT	No	No	No
[14]	Crime Density	Ambiguity Score Based	No	No	Yes
[15]	Crime Article Classification	SVM	No	No	Yes
[16]	Text Classification	Decision Tree	No	No	No
[3]	News Classification	SVM	No	No	No
[1]	Medical Doc. Filtering	SVM	No	No	Yes
[4]	Diseases, text Classification	Decision Tree	No	No	No
[8]	Fake News Detection	BERT	No	No	No
This Work	News Article Analysis	BERT	Yes	Yes	Yes

3 Methodology

The main aim of this work is to analyze agriculture-related news systematically. Moreover, gather all agriculture news together in one place. We can divide the work into five main steps, as shown in Fig. 1. Step A, is the crawling process to collect data from online news sites. Step B is the classification process to check the data between agriculture and non-agriculture. Step C is for the duplication process checking. The stage is to find out if the article was posted previously. After that, the non-duplicate agriculture data will insert into the database. Step D is for sub-classify the agriculture articles to some feature. Finally, Step E recommends articles based on sentiment analysis and time-sensitive content.

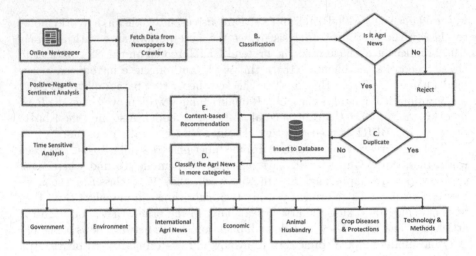

Fig. 1. Flow diagram of the process

3.1   Step A: Fetch Data from Online News Site

We are collecting news articles from several popular online newspapers for our proposed work. However, manually collecting the news by exploring all the newspapers is challenging. This process will take more time and effort, so the process is not an efficient way to gather news. We used a web crawler to fetch data from the online newspaper as a solution. The process needs to follow some steps to crawl data, such as processing online news sites and managing the URL (Uniform Resource Location). The process is an automated way to fetch data from online resources. URL is the main component of the working model. First, we need to analyze the URL in this process, and then data processing will be done in the next step [17]. The process needs to detect which one is an article and which one is an advertisement. By monitoring this, it will ignore all the advertisements from online news sources. Beautifulsoup and Xpath parsing tools collect essential data from online sites. Generally, beautiful soup makes the HTML (Hyper Text Markup Language) more simple and organized to collect the data. To manage the URL scheduler process used to crawl data and find the new links for this process. In a nutshell, at the starting point, the crawler will look for news UI (user interface), fetch the contents, and check for available URLs, including advertisement content, by analyzing the URL. Finally, the process will store all the data captured in the storage. We must run a crawler system daily to gather as much news as in our storage. Moreover, we have set our crawling system to run for ten hours daily. Then we used Cronjob to automate the crawling process daily. We have used ten popular newspapers to collect the agriculture data in our work. Among the ten newspapers, four newspapers are specially for agriculture news. Those newspapers are:

- Times of India

- The Hindu
- NDTV
- News18
- India Today
- Hindustan Times
- Down to Earth
- Krishi Jagran
- Krishak Jagat
- Successful Farming

Here Down to Earth, Krishi Jagran, Krishak Jagat, and Successful Farming mainly focus on agriculture news. Successful farming news site provides most of the new technology news and success stories in agriculture. We have more agriculture and climate-related news of India from Krishi Jagran, Krishak Jagat, and Down to Earth news sites. Using the crawler, we have collected several articles, including agriculture and non-agriculture, from all the mentioned newspapers.

3.2   Step B: Agriculture News Detection

In the database, there is much news, both agriculture and non-agriculture related. For this work, we must first separate the data between agriculture news and non-agriculture news. One way is manually dividing the data into two categories. However, for a large number of data, the process will require more work to identify the agriculture articles manually. The text classification process can be a solution here.

We got 4860 articles and labeled the data between agriculture and non-agriculture news. One thousand articles for both agriculture and non-agriculture data as Table 2. Labeling the data is a prepossessing step for the classification process. The classifier model will be trained on the labeled data to identify agriculture and non-agriculture articles. Many text classifiers are there to do classification. So, here we have tried five different text classification processes: Naïve Bayes, Decision Tree, Random Forest, Support Vector Machine (SVM), and Bidirectional Encoder Representations from Transformers (BERT). The best one will be selected as a classifier model for the following work based on the performance.

Table 2. Data labeling information

Total Data	4860
Labeled Data	2000
Non-agriculture Data	1000
Agriculture Data	1000

The Naïve Bayes model follows Bayes' theorem. The probabilistic model works between two events, agriculture, and non-agriculture, for the proposed work [18]. The decision tree model is based on a tree classification model [19]. We have used the algorithm in our proposed work to get a better classification. Random forest is also known as a good classifier for text data. As well as SVM, it performs very well for classification problems [20]. In our proposed work, we have two levels of data- agriculture, and non-agriculture. TF-IDF (Term frequency-inverse document frequency) method will convert the labeled data into vectors for the SVM working process.

BERT is considered the most successful and efficient deep learning model for (Natural language processing) NLP work. The addition of Semi-Supervised Learning makes BERT more efficient in NLP tasks [5,21]. Pre-training and fine-training are the two phases of BERT model structures [5]. In pre-training, data is unlabeled, and in fine-training, data is labeled. For our case, fine training was used as we labeled our data. The main reason for using BERT in our proposed work is that another classification model works based on some terms, but BERT works based on context. For a large data set, it shows a significant performance. So, BERT can outperform other classifier algorithms. The final model will be selected for this work based on the best classification performance among these five algorithms.

3.3   Step C: Duplication Article Detection

There are multiple news articles in the online newspaper, but there is a possibility of repeating the same news multiple times. One single news site can repeat its news after a few days. The same and another news site can publish the same article with a different title after a few days. In this work, we are going to ignore this duplicate news. Manual checking articles and then removing duplicate articles is not an efficient way. For this purpose, we have used two methods. One is cosine similarity over TF-IDF. First of all, we have to convert the text into a vector [22]. Then a matrix will be created with the information and including the document. In this matrix, the row vector shows the terms, and the column vector shows the documents. In TF-IDF, every term provided a different frequency score based on the score to identify if there were any duplicates or not [23]. For that, a threshold value needs to be selected. Then after getting the vector and frequency, we can check the similarity between the two documents using cosine similarity [24]. Moreover, the second method is a state-of-the-art SBERT (Sentence BERT) model. We have used the pre-trained SBERT model "all-MiniLM-L6-v2" [25], a part of the Sentence Transformers library.

In both cases, we have set the threshold value as 0.6. Any document's frequency score will show as duplicate articles if it exceeds this.

3.4   Step D: Sub-classification Process

After the classification process, we now have only agriculture news. We want to create some subclass of agriculture news. Thus people can visit directly any categories as people want. As a result, we manually labeled our data and identified some amounts of data for various categories. Doing the sub-classification process by seeing all articles is a big challenge. So, here we used a classifier for the sub-classification and trained with our labeled data. We only have agriculture data for various categories in this process, as shown in Fig. 2. The BERT model has been used here for classification because of its superior performance. Here we are classifying the agriculture data into seven other categories such as:

– Government
– Environment
– International
– Economic
– Animal Husbandry
– Crop Diseases & Protections
– Technology & Methods

Here data for subcategories are imbalanced. As a result, the model's performance can not be better. The decision will be biased towards the majority class, as it has more data in this section. On the other hand, the model cannot detect the minority class properly. So accuracy metric will need clarification between classes. In this situation, oversampling is a solution to balance the classes; thus, the model can be trained equally. In this process, there will be added some duplicates values in the minority sections to equal the size of the primary class. Then the model can train equally for all the classes.

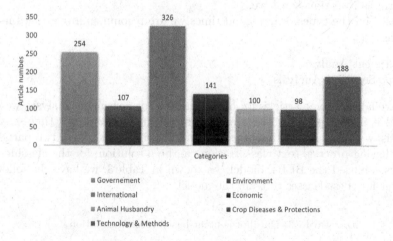

Fig. 2. Data for Sub-classification

After getting the best-classified accuracy and document similarity model, we fixed this in our crawling system. Thus, by following this approach, we do not need to save unnecessary data like duplicate and non-agriculture data. Our approach will directly crawl the data from online resources, and after classifying and checking the duplication process, it will be stored in the database. Moreover, for the following process of the sub-classifier, we have used this same data. As a result, we will have only the agriculture and non-duplicate data and sub-classification feature.

3.5   Step E: Recommendation Process

In the proposed work, There are various kinds of articles. Some articles positively impact people, and some negatively impact people. We are comparing positive and negative articles on agriculture. For example, there are some agriculture articles in which some farmers died, committed suicide, or lost production. Such kinds of articles are considered negative articles. Similarly, some articles are there with a good impression, like a big success, new technology, or winning a prize. These kinds of articles are considered positive articles. In this section, we can call Sentiment analysis based on contents.

Another recommendation is based on time. For example, we can see some news has extreme time limits that we can deliver to people in a time. After that time, the news has no value to people, such kinds of articles we determined as Time sensitive articles. For example, some extreme weather news is considered time-sensitive because this kind of news needs to show instantly. After two or three days, that news is not helpful. On the other hand, we keep more time for some articles. As an example, some success stories and some technology news. This kind of news can keep in our system for multiple days. So, we considered the term as Non-time-sensitive.

Based on the types of news and times we are recommending articles in two sections:

- Sentiment Analysis
- Time Sensitive Analysis

We have labeled some articles for both sections. For Sentiment analysis, we have labeled it in two parts- positive and negative. Furthermore, for time-sensitive analysis, we have labeled time-sensitive and not time-sensitive in two parts. For the following process, text classification can be a solution; for the classification process, we used the BERT model. As shown in Table 3, we have the following data set for these classes to train the model.

Table 3. Data for content-based recommendation

Sentiment Analysis		Time sensitive analysis	
Positive articles	106	Time sensitive	114
Negative articles	110	Not time sensitive	122

4   Result and Analysis

4.1   Agriculture Articles Classification Results

After the classification process, if we analyze the confusion matrix of each algorithm, we can see that the Naive Bayes algorithm provides 93% test accuracy. Furthermore, the Error Rate (ER) for this process is 7%. The Decision tree algorithm provides 95% test accuracy with an error rate of 5%. Random Forest also performed similarly to the Decision tree method. Another classification algorithm, SVM, provides 95% test accuracy and a 5% error rate. However, the BERT transformer algorithm provides the highest test accuracy of 97% with only a 3% error rate, as shown in Table 4. Not only accuracy metric there are other metrics to know any model performance, and those are important also, such as Recall, Precision, and F1 score. Now, in the result for the classification of agriculture and non-agriculture news, we can see that the BERT model has the highest Precision, Recall, and F1 score with 98% as shown in Table 4. Naive Bayes, Decision tree, and Random Forest have the lowest Precision rate of 96%. For Recall, Naive and Decision tree keeps a similar recall rate of 94%, and SVM and Random Forest have 95% in this case. For the F1 score, SVM has the lowest rate of 94%, and BERT shows the highest at 98%. If we compare the BERT algorithm with other text classification algorithms, the BERT algorithm achieves an outstanding performance because it works depending on the context of a text. On the other hand, other algorithms work depending on some critical term or keyword. As a result, other algorithms sometimes show non-agriculture news as agriculture. For example, one article is - "Apple company now producing more new mobile day by day"; this is a piece of non-agriculture news. However, by monitoring the "apple" word, other algorithms configured it as an agriculture article. In that case, the BERT algorithm worked correctly. BERT shows superior performance for big data sets, and we finalized BERT as our system text classification algorithm.

Table 4. Performance Metrics for Agri & Non-agri classification

Performance Metrics	Accuracy	Precision	Recall	F1 Score
SVM	0.95	0.97	0.95	0.94
Naive Bayes	0.93	0.96	0.94	0.95
Decision Tree	0.95	0.96	0.94	0.95
Random Forest	0.95	0.96	0.95	0.96
BERT	0.97	0.98	0.98	0.98

4.2   Duplication Articles Check Results

For the duplication check of the articles, we have set the frequency threshold value of 0.6 in both processes. That means if an article is more than 0.6 similar, it will select that article as a duplicate. Moreover, we have checked the process for the last 20 days of agriculture data. To check the performance, we have created a database of 60 articles. Among these, 30 articles contain similar articles. In this case, for certain news, we took data from several newspapers for the similarity between articles. The other 30 articles are non-similar. Using this, we have checked the performance metrics and got 84% accuracy for the TF-IDF process and 88% accuracy for the Sentence Transformers process. As shown in the Table 5.

Table 5. Results of Duplication Detection Process

Methods	Accuracy
TF-IDF	0.84
Sentence Transformers	0.88

4.3 Sub-classification Results

The following proposed work is to classify the agriculture articles into other categories. We took seven categories to sub-classify the agriculture data. As we have fewer data to label, we got a preliminary result by the BERT classification. For the performance of BERT, we considered this the final classification model for all sections. We got 82% accuracy. And an average Recall, Precision, and F1 score of 81%, as shown in Table 6

Table 6. Performance metrics for Sub-classification Process

Performance Metrics	Rate
Accuracy	0.82
Precision	0.81
Recall	0.81
F1 Score	0.81

Moreover, we can see the details of Recall, Precision, and F1 scores for all the classes separately. Thus we can know which classes are captured most by the model, as shown in Fig. 3. Here, the Crop Diseases and Protections class has the highest Precision value, and the lowest has got by the Government class. For Recall Animal Husbandry class has the highest, and the International class shows the lowest among all. We can measure the overall by the F1 score. Similarly, two classes, Animal Husbandry and Crop Diseases and Protection, achieved the highest and the lowest one, illustrated by the Government class.

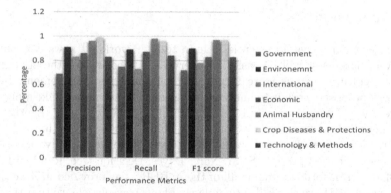

Fig. 3. Performance metrics of all classes of the Sub-classification process

4.4 Content-Based Recommendation Results

We recommend people based on sentiment and time analysis articles in this work. Here, in the sentiment analysis, we took two classes, positive and negative. In the time-sensitive analysis, we have two classes called time-sensitive and not time sensitive. We have used BERT for the classification of this process. Recall, Precision, and F1 score measurement play a vital role in understanding the model performance. Here for both, the process model shows 94% accuracy for sentiment analysis, and for Time-sensitive, it provides 85% accuracy, as shown in Fig. 4. Here we do not want to skip recall and precision value because recall provides accurate positive measurements predicted by the model. We do not want to skip positive sentiment news. Accordingly, do not let go of any negative ones. Also, the precision value provides all positive news, so we do not want to place any urgent news in a time-insensitive class for time-sensitive reasons. Hence, for both process, higher recall and precision value is essential. The work can recommend which articles make positive and negative vibes and which articles need to show in a limited time.

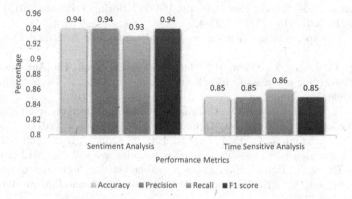

Fig. 4. Results of recommendation based on contents & Time

5   Conclusion

Online news site plays an essential role in today's world. If news can get separately as needed, it will be easier for everyone. Especially for the agricultural area, agriculture news is more helpful for one country's development. In this work, our proposed work is to analyze the agriculture news and make a platform for everyone. Thus, people can get all agriculture-related news together. We have crawled data from ten newspapers in India. Moreover, we have detected the agriculture news from all the news articles. We have used five algorithms for this classification process: SVM, Naive Bayes, Decision tree, Random Forest, and BERT Transformer. Among all of these, BERT has provided 97% accuracy. Furthermore, we have checked the similarity of our news articles so that the same news should not be repeated here. We have finalized the Sentence Transformer model for the duplication process, and the model efficiently worked for this. BERT classifier has been used for the Sub-classification categories of agriculture news in seven other categories. In this case, we got a decent output. Moreover, we recommended news to people based on Sentiment and Time-sensitive analysis. BERT is also used for classification in this recommendation process. In the future, we can add more data for better classification in sub-categories and improve the sub-classification. Also, improve the recommendation system by adding more data and can go for a more efficient method to make the work more valuable. In addition, multilingual languages can be used in the future.

Acknowledgement. Authors acknowledge the grant received from the Department of Science & Technology, Government of India, for the Technology Innovation Hub at the Indian Institute of Technology Ropar in the framework of National Mission on Interdisciplinary Cyber-Physical Systems (NM - ICPS).

References

1. Aggarwal, C.C., Zhai, C.: A survey of text classification algorithms. In: Aggarwal, C., Zhai, C. (eds) Mining Text Data, pp. 163–222. Springer, Boston (2012). https://doi.org/10.1007/978-1-4614-3223-4_6
2. Wight, C.: Speaker classification on general conference talks and byu speeches (2021)
3. Zhang, Y., Dang, Y., Chen, H., Thurmond, M., Larson, C.: Automatic online news monitoring and classification for syndromic surveillance. Decis. Support Syst. **47**(4), 508–517 (2009)
4. Charbuty, B., Abdulazeez, A.: Classification based on decision tree algorithm for machine learning. J. Appl. Sci. Technol. Trends **2**(01), 20–28 (2021)
5. González-Carvajal, S., Garrido-Merchán, E.C.: Comparing bert against traditional machine learning text classification. arXiv preprint arXiv:2005.13012 (2020)
6. He, K., Zhang, X., Ren, S., Sun, J.: Deep residual learning for image recognition. In: Proceedings of the IEEE Conference on Computer Vision and Pattern Recognition, pp. 770–778 (2016)
7. Adhikari, A., Ram, E., Tang, R., Lin, J.: Docbert: bert for document classification. arXiv preprint arXiv:1904.08398 (2019)

8. Qasim, R., Bangyal, W.H., Alqarni, M.A., Almazroi, A.A.: A fine-tuned bert-based transfer learning approach for text classification. J. Healthcare Eng. 2022 (2022)
9. Song, Z., Xie, Y., Huang, W., Wang, H.: Classification of traditional Chinese medicine cases based on character-level bert and deep learning. In 2019 IEEE 8th Joint International Information Technology and Artificial Intelligence Conference (ITAIC), pp. 1383–1387. IEEE (2019)
10. Kumar, A., Gupta, P., Balan, R., Neti, L.B.M., Malapati, A.: Bert based semi-supervised hybrid approach for aspect and sentiment classification. Neural Process. Lett. 53(6), 4207–4224 (2021)
11. Munikar, M., Shakya, S., Shrestha, A.: Fine-grained sentiment classification using bert. In: 2019 Artificial Intelligence for Transforming Business and Society (AITB), vol. 1, pp. 1–5. IEEE (2019)
12. Garg, S., Ramakrishnan, G.: Bae: bert-based adversarial examples for text classification. arXiv preprint arXiv:2004.01970 (2020)
13. Hwang, S., Kim, D.: Bert-based classification model for korean documents. J. Soc. e-Business Stud. 25(1) (2020)
14. Gupta, S.K., Shekhar, S., Goel, N., Saini, M.: An end-to-end framework for dynamic crime profiling of places. In: Smart Cities, pp. 113–132. CRC Press (2022)
15. Jayaweera, I., Sajeewa, C., Liyanage, S., Wijewardane, T., Perera, I., Wijayasiri, A.: Crime analytics: Analysis of crimes through newspaper articles. In: 2015 Moratuwa Engineering Research Conference (MERCon), pp. 277–282. IEEE (2015)
16. Sharma, V., Kulshreshtha, R., Singh, P., Agrawal, N., Kumar, A.: Analyzing newspaper crime reports for identification of safe transit paths. In: Proceedings of the 2015 Conference of the North American Chapter of the Association for Computational Linguistics: Student Research Workshop, pp. 17–24 (2015)
17. Lu, M., Wen, S., Xiao, Y., Tian, P., Wang, F.: The design and implementation of configurable news collection system based on web crawler. In: 2017 3rd IEEE International Conference on Computer and Communications (ICCC), pp. 2812–2816. IEEE (2017)
18. Ming Leung, K.: Naive bayesian classifier. Polytechnic University Department of Computer Science/Finance and Risk Engineering, 2007, 123–156 (2007)
19. Swain, P.H., Hauska, H.: The decision tree classifier: design and potential. IEEE Trans. Geosci. Electron. 15(3), 142–147 (1977)
20. Zhang, Y.: Support vector machine classification algorithm and its application. In: Liu, C., Wang, L., Yang, A. (eds.) ICICA 2012. CCIS, vol. 308, pp. 179–186. Springer, Heidelberg (2012). https://doi.org/10.1007/978-3-642-34041-3_27
21. Farzindar, A., Inkpen, D.: Natural language processing for social media. Synthesis Lectures Hum. Lang. Technol. 8(2), 1–166 (2015)
22. Alodadi, M., Janeja, V.P.: Similarity in patient support forums using TF-IDF and cosine similarity metrics. In: 2015 International Conference on Healthcare Informatics, pp. 521–522. IEEE (2015)
23. Manning, C.D., Raghavan, P., Schutze, H.: Introduction to information retrieval, vol. 1. Cambridge University Press, Cambridge (2008)
24. Salton, G.: Automatic text processing: The transformation, analysis, and retrieval of. Reading: Addison-Wesley, 169 (1989)
25. Technische Universität Darmstadt Nils Reimers. Pretrained Models - Sentence-Transformers documentation. https://www.sbert.net/docs/pretrained_models.html. Accessed 12 Dec 2022

Intelligent Chatbot Assistant in Agriculture Domain

Rahul Biswas[✉] and Neeraj Goel

Department of Computer Science and Engineering, Indian Institute of Technology
Ropar, Punjab, India
{2021csm1013,neeraj}@iitrpr.ac.in

Abstract. Agriculture is known as the economic game changer of India. It is the primary driver of GDP growth because of India's robust agricultural industry, and proper knowledge about agriculture techniques help increase crop yield. So, answering the different types of crop-related queries is essential. We proposed the intelligent chatbot application in the agriculture domain so that farmers can get the correct information about farming practices. Our system is farmer-friendly and capable enough to instantly answer farm-related queries from the knowledge base, such as plant protection, fertilizer uses, government schemes, and many others. We used the agriculture-related data in question-answer format and implemented the pre-trained model of the Sentence-Transformer approach to answer providing. We also deployed the TF-IDF and Bag-of-Words method but achieved a reasonable accuracy rate for the test data in the sentence transformer pre-trained model. With the help of API services, our system also shows the crop's latest mandi (market) rate and current weather information. So, the proposed chatbot system will keep the contribution for farmer's cost savings. Overall, our chatbot system is straightforward and more efficient for the farmer to make better decisions.

Keywords: Sentence Transformer Pre-Trained Model · TF-IDF · Bag of Words · Pegasus Model · Mandi(Market) Rate API · Cosine Similarity · Weather API

1 Introduction

In Asia, most countries depend on agriculture to fulfill the food demand of the people, and agriculture is a significant contributor to the country's increasing productivity. In India, nearly 60% of the population works in agriculture [1], and lots of people from rural areas are farming for their livelihood as their occupation [2]. However, most Indian farmers do their farming process in their traditional method, and because of the technological and other communication gaps, they are unknown of the latest farming information. Most of the time, they are unable to solve farming-related problems. As a result, they experience limited crop growth, which causes them to incur losses.

M. K. Saini et al. (Eds.): ICA 2023, CCIS 1866, pp. 180–194, 2023.
https://doi.org/10.1007/978-3-031-43605-5_14

So, to solve the agriculture sector issues, the government already take many necessary steps. The farming authority also organizes several programs for farmers, like Front Line Demonstrations with the help of a network of Krishi Vigyan Kendras in every district, Rashtriya Krishi Vikas Yojana, and other services [3]. Moreover, we can see many existing services for farmers, like the Kisan calling center, eNam, and farmers portal.

The Ministry of Agriculture and Farmers Welfare is continuing the Kisan call center (KCC) to facilitate efficient communication of farmer's queries in various languages. However, due to many incoming calls, it is sometimes challenging to attend to all of them promptly, and users have to wait a long time to ask their queries. As a result, users need help getting the solution to their query. Besides, Agriculture Ministry has also introduced a web-based service called the Farmers Portal, which offers a comprehensive range of information and services to farmers in one convenient site. The main feature of this digital platform is agri-Advisory, crop management, animal husbandry management, and so on [4]. Despite its benefits, the Farmers Portal has some limitations, including fixed support, and users may need to have the technical knowledge to navigate the platform. These challenges may be complex for some users to access required agricultural information. Moreover, we can see the eNAM - National Agriculture Market platform for selling the farmer produce, the Kisan Suvidha application, which integrates all the farm-related services.

Most of the existing works used the traditional approach or pre-implemented platform to deploy the farmer chatbot applications, which may be less suitable for a large number of chatbot knowledge. We can also see the most recently introduced technique, CHATGPT, which can answer all types of questions. However, in the agriculture domain, some factors also limit this application. We introduced an intelligent chatbot system for farmers to get suitable solutions for various farming practice queries. We aim to enhance farmer operations and eventually support the agriculture industry's expansion and sustainability. One of the biggest challenges of deploying an agriculture chatbot application is ensuring reliable and efficient data. The collected data for the agriculture chatbot application must be relevant, accurate, and up-to-date to provide the best possible answer and support. To overcome this challenge, we collected data from KCC and focused on manual data collection, knowledge models, and real-time API information. In our chatbot, users can get answers like plant protection, cultural practices, fertilizer uses, water management, nutrient management, weed management, field preparation, and varieties. Our system also provides the current market price of different crops in India and real-time weather information through the API system. Our system also supports the Hindi language besides English. We developed our application in the Indian context, so we consider the Hindi language as it is the official language of India. We deployed the latest AI technique and built a user-friendly system for non-technical individuals. Our chatbot is convenient and provides customized information about farming-related user queries.

The paper organizes as follows: Sects. 2 and 3 discuss the related works and the proposed methodology. Section 4 explains the dataset collection and pre-possessing. Section 5 represents the dataset analysis. Sections 6 and 7 describe the approaches and analyze the test result. Finally, Sect. 8 concludes the overall work.

2   Related Works

Bhardwaj et al. [5] proposed the farmer-assistive chatbot using the KNN and the sequence-to-sequence model for the answer generation that is multi-linguistic supported. Arora et al. [6] implemented the RNN seq2seq technique for the conversational system using the KCC and web scraping data. They also used the CNN approach to detect crop disease, and the weather information will come through the API services. Niranjan et al. [7] discussed the overall survey on the chatbot application. They described the shortcomings of the various technique and suggested the seq2seq as a good solution in the case of the chatbot assistant.

Gounder et al. [8] and Mohapatra et al. [9] implemented the TF-IDF approach to generate the vector and used the cosine similarity method to find the similarities of the question from the KCC dataset. Besides, they [8] considered the lemmatization process instead of the steaming process for the data preprocessing and deployed the application in Android.

Nayak et al. [10] introduced the agriculture expert chatbot application using the chatterbot library, where they collected agriculture-related data in question-answer format. They used the Levenshtein distance and the best-match approach to answer the user queries correctly. Here, users can also ask their query through voice and get agro-expert support in case of chatbot is unable to answer.

Jain et al. [11] deployed the FarmChat android application with audio and (audio+text) and compared the output efficiency in their study. Here, they considered not only the KCC data but also collected information from local cultivators and agriculturalists so that the application could answer farm-related questions about potato crops easily. Thatipelli et al. [12] designed an agriculture robot with different types of sensors for collecting data from the farmland and storing it in the IBM cloud for real-time analysis. Besides, they also deployed the chatbot based on the collected data, which suggests a suitable fertilizer considering different parameters and detects plant diseases. Momaya et al. [13] also introduced the farmer assistive chatbot, named "Krushi" using the RASA X framework, which is able to provide almost 70% of query answers.

Gunawan et al. [14] proposed an IoT - based chatbot application. With the help of an IoT device, it senses the field humidity, temperature, and soil moisture and sends the sensor data to the database regular basis. The chatbot uses this data to provide real-time information and help urban farmers to improve agriculture's efficiency. Mostaco et al. [15] proposed the AgronomoBot, which can also answer the farmer's queries from the agriculture wireless sensor network data. The data will come from the WSN database and deploy the application in the telegram for the user interface. Kiruthika et al. [16] proposed a quite different chatbot system for the consumer and the farmers, where a farmer can sell their crops at a suitable price and get more profit. Their proposed approach's main functionalities are collecting the farmer details and crop-related information, recommending farmers as the user needs, and constantly updating the farmer crops details.

Shao et al. [17] proposed a neural network-based transformer to solve the tasks of answer selection in the question-and-answer systems. They implemented the (BiLSTM) bidirectional long short-term memory, transformer-based, and prioritized sentence embedding tasks. Here, Ngai et al. [18] deployed different transformer-based models like BERT, ALBERT, and T5 for Question Answering (QA) systems on COVID-19. In the result analysis, they achieved the effective F1 score for the BERT and a better exact match for the ALBERT technique. Soldaini et al. [19] proposed another answer selection approach, named cascade transformer. On several benchmark datasets, they evaluated the Cascade Transformer, and the experiment's outcomes indicate that the Cascade Transformer is a promising method for choosing answer sentences in QA applications.

3   Proposed Methodology

We proposed an effective chatbot application to answer farm-related queries using state-of-the-art technology. For the agriculture chatbot application, the dataset plays a significant role. We combined various sources of data in the question-answer structure. We imported the necessary library as model requirements, preprocessed the data, and loaded the pre-trained model. Here, the sentence transformer pre-trained model encoded the input query and the knowledge base questions and computed similarity scores using the cosine similarity approach. We set a threshold value based on the chatbot's performance. Moreover, if the user wants to know the Weather Information or Mandi Rate, our chatbot will display the information through the API services regarding their search. For the weather condition, users need to input the location name, like city, and to get mandi rates with more information; they must search by crop name. As we already said, our system is user-friendly, meaning it will also answer the query in Hindi. We deployed the google translate API for language translation. So, when users ask a query in Hindi, it will translate into English, find a similar question with an answer, and deliver it in Hindi. In the chatbot, users may ask a question in different ways. So, to analyze the results, we used the Pegasus model to generate paraphrases of the test questions and checked whether the model identified the correct answer or not. We implemented different approaches but found that the sentence-transformer pre-trained model provided the best results for the test dataset. In Fig. 1, we can see our proposed work's architecture.

4   Dataset Collection and Preprocessing

The utilization of datasets is a crucial part of developing question-answering chatbots. We collected the dataset from different sources and prepossessed it to build our farmer chatbot system. In this section, we will discuss it more elaborately.

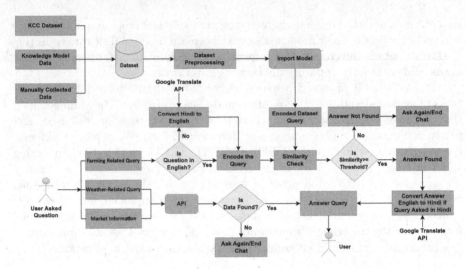

Fig. 1. Proposed Architecture

4.1   KCC (Kisan Calling Center)

Earlier, we discussed the Kisan calling center, a help center facility for the farmer. A total of 21 Kisan call center locations exist in different states of India. This kind of service helps the farmer acquire more knowledge about farming. For the farmer chatbot application in Indian Region, the KCC dataset is essential. This open-source dataset is available on the government website [20] (data.gov.in). So we collected the most recent KCC data of Punjab, Tamil Nadu, and Maharastra for the different districts. We have 65.11% for the Punjab state, Maharastra 22.39%, Tamil Nadu 12.48%, and more than fifteen thousand total KCC raw data. The dataset has nine features: Seasons, Sectors, Crops, Query Type, Query Text, Answer, StateName, District Name, and CreatedOn. However, in this case, we considered only the valuable feature that contributes most to the provided answer in our chatbot application.

- **Category:** This column represents the category of the query asked by the users, like vegetables, cereals, etc.
- **Crop:** This column represents the specific crop mentioned in the user's query.
- **Query Type:** This column represents the type of query asked by the users, like plant protection, varieties, etc.
- **Query Text:** This column represents the query asked by the users.
- **KCC Answer:** This column represents the answer given by the Kisan Calling Center representative to the user's query.

Data Preprocessing. After the dataset collection, it is essential to preprocess the data. The KCC dataset was in JSON format, and we converted it into CSV format to understand the data more efficiently. We merged all the district-wise KCC collected data and evaluated it for further procedure. After analyzing the KCC data, we found lots of duplicate data, spelling mistakes, incomplete sentences, etc. We followed the remaining steps to preprocess the KCC datasets.

- **Used Google Translate API:** The query answers of the KCC dataset were in the local language. To build our system, we considered only the English language. So, we translated the local language to English for better understanding and efficiency in working with the dataset.
- **Dropped the Weather-Related Rows:** Most KCC dataset queries were related to weather conditions. We dropped the weather-related query almost 30% of the total KCC data because the user will get the answer through the weather API services.
- **Dropped the Contact Number Asked Rows:** We found 7% data related to asking for the contact number details of Krishi Vigyan Kendra, agriculture institution, and nodal officer details. However, there needed to be more data, and besides, most of the government scheme-related query answers were the nearest agriculture center contact number. This kind of answer may not be helpful to the user. So, we did not count it because we collected the contact details manually for all asked places in the KCC dataset query and added more government scheme data with suitable answers.
- **Dropped the Duplicates Rows:** We also found lots of duplicate data, which is 40% in the KCC datasets. So, we dropped the duplicates.
- **Dropped Animal and Flowers Related Data:** We did not count the animal and flower-related data in our KCC data consideration. However, in the future, we plan to incorporate this information to enhance the knowledge of our chatbot.
- **Dropped the Farming Fair Information:** In analyzing the Kisan Calling Center dataset, we came across numerous queries related to agriculture fair information. We realized that the queries were referring to fairs holding dates that took place in the past years, and the answer was no longer valid. Therefore, we excluded such queries from our dataset.
- **Dropped the Market Information of Crop Price:** We dropped the market information-related data by almost 1.48% because our chatbot system will provide the information regarding the market crop price through the API services.
- **Others:** After completing the aforementioned steps, we noticed that the dataset had many incomplete answers, inappropriate information, null value rows, and insufficient crop data. Additionally, many repeated questions with different structures could cause problems evaluating the model's performance. So we removed these kinds of information and decided to focus on a limited number of crops with sufficient information available to improve the dataset's overall quality and ensure the effectiveness of the chatbot application.

After the KCC data cleaning, it is quite understandable that the considered data is not sufficient enough for our proposed chatbot application. The following sections will discuss other data collection procedures and API data.

4.2 API Data

In the agriculture sector, knowing about the recent update on weather conditions and the crop price in a different market is essential so that farmers can make immediate decisions. As discussed earlier, we aim to bring all the farming-related functionalities into our chatbot application. So for that, we also deployed the following:

- **Weather Condition**
- **Mandi Rates of Various Commodities**

With the help of open weather API, we displayed the real-time weather parameter like temperature, humidity, weather report, and wind speed of the searching place [21]. It will help farmers with efficient planning and better resource management. The second is the "Current Daily Price of Various Commodities from Various Markets or Mandi." The API services are available on the [20] (data.gov.in) website, which is open-source for all. This API provides ten features like state, district, market Name, crop name, variety, grade, data arrival dates, crops minimum, maximum, and model price (Rs./Quintal). In our chatbot application, users only have to input the commodities names; according to that, our chatbot will show the Mandi Rates with state, district, and Market names. Here, we also deployed the spell-checker functionalities in the commodities search to keep in mind that if the user makes mistakes in writing the crop name, it will automatically take the correct spelling.

4.3 Knowledge Model

In the farmer chatbot application, the completeness of data plays a significant role because it will help to enlarge the knowledge and enable the application to answer different crop-related questions. We all know that most people in India depend on rice and wheat crops. However, we found a lack of information on disease protection, weed management, and water management for these crops in the KCC dataset. To address this gap and enlarge the chatbot's knowledge, we collected data from agricultural experts on rice, wheat, and maize crops. Although the data was not in a question-and-answer format, it contained valuable information on crop varieties, sowing times, disease symptoms, protection procedures, nutrient management, and many more. Overall, we found the highest number of rice data, and maize is in the last position, where rice has 106 varieties, wheat 64, and maize 25. For each variety, there are also sowing time details and features. We also noticed the highest number of disease management data for rice and pest management for wheat. It is necessary to the conceptual representation of the data and its relationships to understand the data completeness or required information. In that case, the knowledge model is more adaptable and scalable. It provides a visual understanding and clearly and consistently represents structured data experience. The flexibility of the knowledge models helps to accumulate information in an organized manner and create better decisions. We took inspiration from the Agropedia source [22] to design the knowledge model. In this work, we developed the knowledge model of agriculture for rice, wheat, and maize crops. These models focused on different concepts: nutrient management, environmental requirement, varieties, field preparation, sowing time, soil management, fertilizers, water management, weed management, pest management, and diseases management. So these knowledge models helped us to comprehend data completeness, identify the missing information, and understand the relationship. We converted this data into a question-answer format that can enhance the efficiency of chatbot applications and contribute towards farmer's adoption of successful as well as productive agricultural practices (Fig. 2).

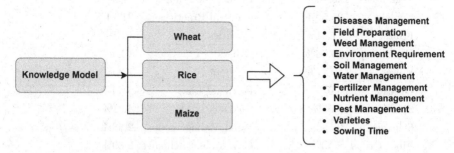

Fig. 2. Different Concepts of Rice, Wheat, and Maize Knowledge Models

4.4 Manually Data Collection

To make our chatbot application more efficient and a better resource, we took help from various governments and farming-related sources [23–29] for required query type data. So we manually created different questions regarding the required data and stored the information in the answer structure. To cater tó the farmer's diverse information needs, we prioritized the various query types for each crop, ensuring that a range of information will be available. We also included frequently asked questions about government schemes, such as how to apply, required documents, benefits, and other necessary information. Moreover, in the previous section, we discussed many queries about the Krishi Vigyan Kendra, agriculture institution, and Nodal officer contact details. So, we added the contact number of Krishi Vigyan Kendra all over India with senior scientist details, more than 60 agriculture institutions, and nodal officer information. Most of the contact details data was in the tables layout in the government sources. We arranged the collected information in the question-answer structure to help the chatbot system provide a suitable answer to the user.

5 Dataset Analysis

After collecting the data, we have five columns: Category, Items, Query Type, Query, and Answer. Here the item part includes the different crop names and the government scheme name.

Table 1. Analysis of Crop's Category and Query Type

Category	Total Data (%)	Query Type	Total Data (%)
Vegetables	25.84%	Plant Protection	27.38%
Cereals	25.50%	Cultural Practices	21.14%
Fruits	11.31%	Varieties	15.16%
Others	10.58%	Others	10.58%
Pulses	10.17%	Nutrient	7.36%
Oilseeds	6.07%	Fertilizer Uses	6.49%
Spices	5.55%	Seeds	4.61%
Schemes	2.59%	Weed Management	2.38%
Fiber Crops	2.30%	Government Schemes	2.69%

Table 1 shows the category column information and the count of different agricultural query types. The largest segment of the table represents the vegetable category, almost 25.84 % of the total count-the next largest segment for the Cereals category, followed by the fruits. The remaining categories, such as Schemes, Pulses, and others, would occupy smaller segments proportional to their counts. In Table 1, the most common query type is plant protection, with a count of 27.38%, followed by cultural practices, with 21.14% queries. These two query types represent over 49% of the total queries. The next is Varieties, with a count of 15.16%, followed by other query types, with an exact count of 10.58%. The dataset contains 496 contact details for KVK centers and information about various government schemes, such as Pradhan Mantri KISAN Samman Nidhi Yojana and the PM Kisan credit card scheme. Moreover, the dataset also adds the contact details of agriculture institutions, universities, and Nodal Officers.

Besides, we have a vast number of rice data, with a 14% count representing the highest crop data. The next largest crop is Wheat, with a count of 6.35%, followed by maize and Bengal gram. Other crops, such as Potato, Brinjals, Tomato, Garlic, and Cauliflower, are also in there but to a lesser extent. The remaining crops, such as Coriander, Mustard, Chillies, Turmeric, and others, represent a smaller proportion of the total crops.

The main challenge in the farmer's question-answer chatbot application is collecting accurate and suitable data. Besides the Kisan calling center data, we also included the missing information from scratch, making the system more efficient for the farmer. Overall, there are more than six thousand total data. The analysis shows that the chatbot application has sufficient query type knowledge for the limited crop, government scheme, and contact number details to enhance the chatbot's ability to understand and respond accurately to user queries.

6   Models

For deployed the chatbot application, we used three approaches: Bag-of-Words, TF-IDF, and a Pre-trained model of the sentence transformer. These methods are famous and can see the contribution to NLP tasks, including intent classification and entity recognition. Bag-of-words is a popular approach for representing text data as numerical vectors. Depending on the specific task, it is traditional and valuable. In our chatbot application, we used the BoW to generate the feature vector of the questions. Secondly, the widely used method in NLP is called TF-IDF (Term Frequency-Inverse Document Frequency). Unlike BoW, the TF-IDF prioritizes the significant word. In our chatbot, we also used the TF-IDF technique to generate the feature vector of the queries.

Lastly, Sentence Transformer is the latest approach in NLP that provides pre-trained models for various uses. Here, we used the quite effective, more petite, faster pre-trained model "all-MiniLM-L12-v2" and retained a high level of accuracy as well as good performance [30]. The models pre-trained on extensive textual data to enhance performance on particular tasks. We used the pre-trained model for encoding text into vector representations, where generating sentence encoding involves several steps. The model first tokenizes the input sentence by splitting it into individual words and integrates pooling and other processes in the encoding phase. It can understand the connections between phrases, words, and their context in a sentence through the pre-training procedure. The pre-trained sentence transformer model encoded knowledge-base questions and the user-asking query. The highest cosine similarity score counted to determine the most similar question from the knowledge base for asking the query. Through the testing part, we noticed that this pre-trained model also performed better in the case of synonym words in a sentence and took less time to provide the answer.

7   Result and Observations

After implementing the chatbot application, it is mandatory to check whether it will provide the correct answer. So, we randomly selected unique 100 and 200 questions from our original datasets as test data with the correct answer. After that, we deployed the pre-trained pegasus model for both test datasets to generate the paraphrase of these questions to analyze the model performance. Pegasus is a transformer-based language model used for the NLP tasks like summarization [31]. We used the pre-trained pegasus model to produce paraphrased phrases based on its internal understanding of the input text. Human evaluation is essential in machine-generated text, so we manually evaluated the pre-trained pegasus model [32,33]. This process is time-consuming but flexible.

It also helped us understand the paraphrased sentence's meaning and whether it was relevant to the original sentence or not. We checked different facts in each of the paraphrased sentences, such as Meaning, Fluency, and Relevance. We observed that most of the paraphrased questions semantically connected with the original dataset questions, and some were incomplete, such as the actual questions that correctly defined the crop variety, like "Information Regarding Soybean variety SL 958 (2014)?" but in the paraphrased sentence "Is there any information soybean variety SL 958?" and for asking contact number or address of a place like Kisan Vigyan Kendra, the paraphrased question missed the state or district name of KVK. We also counted these questions because, in the chatbot application, users can ask queries in different ways. So this helped us evaluate the system's effectiveness.

We considered the accuracy, average cosine similarity score, and the threshold value for evaluating all three similarity models. The threshold performs as a required minimum similarity score before a question can be considered a match. We compared the predicted answer for each question with the actual answer for accuracy count. A higher accuracy signifies that the system works effectively, identifying the questions most similar to each other and providing the correct answers for most inquiries.

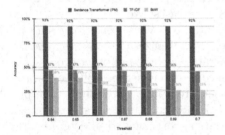
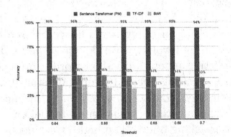

Fig. 3. Test Accuracy of 100 Questions for different threshold values

Fig. 4. Test Accuracy of 200 Questions for different threshold values

Figures 3 and 4 show the accuracy of the Sentence Transformer pre-trained model, TF-IDF, and BoW for 100 and 200 test questions. Here, we considered the different threshold values (0.64 to 0.70) for the model's performance analysis. We defined the threshold value in our chatbot application based on the excellent output. The Sentence Transformer pre-trained model (PM) has the highest accuracy of 93% for a 0.64 threshold value. The TF-IDF model has an accuracy range from 47% to 45%. This model is less accurate than the Sentence Transformer (PM) model but performs better than the BoW model. The BoW model has an accuracy range from 39% to 26%. This one is the least accurate among the three models and outperformed compared to the others. As seen from the Fig. 4 for 200 questions, the Sentence Transformer (PM) has the highest accuracy, followed by TF-IDF and BoW. The accuracy of the models decreases as the threshold increases. At the threshold of 0.64, the Sentence Transformer (PM) has an accuracy of 96%, TF-IDF has 46%, and BoW has 36%. When the threshold goes to 0.7, all approaches' accuracy decreases. The accuracy results may vary based on the dataset and the asked questions. The threshold value is also an essential factor in determining the model's accuracy, and a higher threshold value can result in lower accuracy.

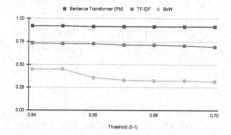
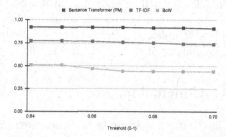

Fig. 5. Average Cosine Similarity Score for 100 Questions for All Three Models

Fig. 6. Average Cosine Similarity Score for 200 Questions for All Three Models

In NLP, the similarity-matching approach is well-known and determines the similarity between two texts or documents. We calculated the average cosine similarity by summing up the similarity scores of the highest-scoring question, which also satisfied the threshold value, and then dividing that by the number of total questions. For the 100 questions, in Fig. 5, the Sentence Transformer method consistently has a better cosine similarity value than the other two methods, with a score of 0.9139 to 0.9205 across all threshold values. The TF-IDF method scored from 0.693 to 0.7337, while the BoW method scored from 0.315 to 0.447. For the 200 questions, in Fig. 6, we can see that the Sentence Transformer model has the highest average cosine similarity score, followed by the TF-IDF and BoW models. At a threshold of 0.64, the Sentence Transformer model has an average cosine similarity score of 0.9211, which is higher than the scores for the TF-IDF model 0.7725 and the BoW model 0.5095. Each model's average cosine similarity score decreases as the threshold value increases. At a threshold of 0.7, the Sentence Transformer model still has the highest average cosine similarity score. Overall, we also find the effectiveness of the Sentence Transformer pre-trained model in this case.

Asked Question: Can you tell me about the potato seed treatment?
Question find from Database: Information Regarding seed treatment of potato?
Answer: For seed treatment in potatoes, 2.5 ml of Monseron per liter of water should be mixed and the seeds should be immersed in this solution for 10 minutes.

Example 1. Q&A of the Chatbot Application

Here, we can see the question-answering process of the Sentence Transformer pre-trained model. The sentence transformer pre-trained model provides adequate performance in the question-answer system. We got 96% accuracy for the 200 and 93% for test question 100 for the sentence transformer (PM), which is the highest compared to the others. Our chatbot is limited in knowledge, meaning it can only answer the database question and supports only Hindi and English. We will enlarge our database to answer all agriculture-related questions in the future and make it available in other languages. Our chatbot assistant is less complex and helps farmers maximize agricultural yields that lead to the development of the agriculture industry, and improve efficiency and profitability.

8   Conclusion

The chatbot system is the most extensive creation in Artificial Intelligence because it solves the user problem quickly and positively impacts the world. Our proposed agricultural chatbot system enables immediate responses to farming-related inquiries in Hindi and English. We used the Sentence Transformer pre-trained method, a state-of-the-art technique in NLP, because of large-scale pre-training and high performance. We obtained a higher accuracy rate in the pre-trained model compared to the other two methods. The proposed method will increase farmer's knowledge and help them make informed decisions about their crop management and operations by providing an accessible, user-friendly system with reliable information. Through the API system, the user will receive real-time responses, which will help them stay more current on market prices and weather information. In the future, we will increase the chatbot's knowledge and add more functionalities like suggesting the crop advisor details who is more supportive regarding the queries. We will also add the query suggestions method depending on the user asking the question and make the system robust.

Acknowledgement. Authors acknowledge the grant received from the Department of Science & Technology, Government of India, for the Technology Innovation Hub at the Indian Institute of Technology Ropar in the framework of National Mission on Interdisciplinary Cyber-Physical Systems (NM - ICPS).

References

1. Statista Research Department Aprajita Minhas. Topic: Agriculture in India (2023). https://www.statista.com/topics/4868/agricultural-sector-in-india/#topicOverview. Accessed 30 Jan 2023
2. Adi(Alternative Development Initiative) International. Topic: Agriculture in India. https://www.adi-international.org/agriculture-rural-development/. Accessed 03 Feb 2023
3. Ministry of Agriculture and Farmers Welfare. Increasing Knowledge and Awareness Among Farmers to Enhance the Production and Productivity (2015). https://pib.gov.in/newsite/PrintRelease.aspx?relid=124565. Accessed 03 Feb 2023

4. Department of Agriculture, Cooperation, and Farmers Welfare. Farmer's Portal. https://farmer.gov.in/FarmerHome.aspx. Accessed 05 Feb 2023
5. Bhardwaj, T., Deshpande, P., Murke, T., Deshpande, S., Deshpande, K.: Farmer-assistive chatbot in Indian context using learning techniques. In: Mahalle, P.N., Shinde, G.R., Dey, N., Hassanien, A.E. (eds.) Security Issues and Privacy Threats in Smart Ubiquitous Computing. SSDC, vol. 341, pp. 239–246. Springer, Singapore (2021). https://doi.org/10.1007/978-981-33-4996-4_16
6. Arora, B., Chaudhary, D.S., Satsangi, M., Yadav, M., Singh, L., Sudhish, P.S.: Agribot: a natural language generative neural networks engine for agricultural applications. In: 2020 International Conference on Contemporary Computing and Applications (IC3A), pp. 28–33. IEEE (2020)
7. Niranjan, P.Y., Rajpurohit, V.S., Malgi, R.: A survey on chat-bot system for agriculture domain. In: 2019 1st International Conference on Advances in Information Technology (ICAIT), pp. 99–103. IEEE (2019)
8. Gounder, S., Patil, M., Rokade, V., More, N.: Agrobot: an agricultural advancement to enable smart farm services using NLP. J. Emerg. Technol. Innov. Res. (2021)
9. Mohapatra, S.K., Upadhyay, A.: Using tf-idf on kisan call centre dataset for obtaining query answers. In: 2018 International Conference on Communication, Computing and Internet of Things (IC3IoT), pp. 479–482. IEEE (2018)
10. Nayak, V., Sowmya, N.H., et al.: Agroxpert-farmer assistant. Global Trans. Proc. **2**(2), 506–512 (2021)
11. Jain, M., Kumar, P., Bhansali, I., Liao, Q.V., Truong, K., Patel, S.: Farmchat: a conversational agent to answer farmer queries. In: Proceedings of the ACM on Interactive, Mobile, Wearable and Ubiquitous Technologies, 2(4), pp. 1–22 (2018)
12. Thatipelli, P., Sujatha, R.: Smart agricultural robot with real-time data analysis using IBM Watson cloud platform. In: Baredar, P.V., Tangellapalli, S., Solanki, C.S. (eds.) Advances in Clean Energy Technologies. SPE, pp. 415–427. Springer, Singapore (2021). https://doi.org/10.1007/978-981-16-0235-1_33
13. Momaya, M., Khanna, A., Sadavarte, J., Sankhe, M.: Krushi-the farmer chatbot. In 2021 International Conference on Communication information and Computing Technology (ICCICT), pp. 1–6. IEEE (2021)
14. Gunawan, R., Taufik, I., Mulyana, E., Kurahman, O.T., Ramdhani, M.A., et al.: Chatbot application on internet of things (IoT) to support smart urban agriculture. In: 2019 IEEE 5th International Conference on Wireless and Telematics (ICWT), pp. 1–6. IEEE (2019)
15. Mostaco, G.M., De Souza, I.R.C., Campos, L.B., Cugnasca, C.E.: Agronomobot: a smart answering chatbot applied to agricultural sensor networks. In: 14th International Conference on Precision Agriculture, vol. 24, pp. 1–13 (2018)
16. Kiruthika, U., Kanaga Suba Raja, S., Balaji, V., Raman, C.J.: E-agriculture for direct marketing of food crops using chatbots. In: 2020 International Conference on Power, Energy, Control and Transmission Systems (ICPECTS), pp. 1–4. IEEE (2020)
17. Shao, T., Guo, Y., Chen, H., Hao, Z.: Transformer-based neural network for answer selection in question answering. IEEE Access **7**, 26146–26156 (2019)
18. Ngai, H., Park, Y., Chen, J., Parsapoor, M.: Transformer-based models for question answering on covid19. arXiv preprint arXiv:2101.11432 (2021)
19. Soldaini, L., Moschitti, A.: The cascade transformer: an application for efficient answer sentence selection. arXiv preprint arXiv:2005.02534 (2020)

20. Ministry of Electronics National Informatics Centre (NIC) and Information Technology. Open Government Data (OGD) Platform India. https://data.gov.in/. Accessed 05 Feb 2023

21. Open Weather Map. https://openweathermap.org/. Accessed 05 Feb 2023

22. Agropedia. Knowledge Models. https://agropedia.iitk.ac.in/content/knowledge-models-0. Accessed 03 Feb 2023

23. Apni Kheti - Empowering Rural India Digitally. https://www.apnikheti.com/en/pn/home. Accessed 18 Dec 2022

24. Punjab INDIA Punjab Agricultural University (PAU) Ludhiana. https://www.pau.edu/

25. New Delhi ICAR-Indian Agricultural Research Institute. https://iari.res.in/index.php/en/iari-new-delhi

26. IIWBR - Indian Institute of Wheat and Barley Research. https://iiwbr.icar.gov.in/

27. Jal Shakti Abhiyan. https://ejalshakti.gov.in/jsa/

28. Direct Benefit Transfer In Agriculture Mechanization. https://agrimachinery.nic.in/Master/User/StateWiseNoduleOfficerReport

29. Vikaspedia Domains. https://vikaspedia.in/agriculture

30. Technische Universität Darmstadt Nils Reimers. Pretrained Models - Sentence-Transformers documentation. https://www.sbert.net/docs/pretrained_models.html. Accessed 12 Dec 2022

31. Pegasus. https://huggingface.co/docs/transformers/model_doc/pegasus. Accessed 03 Feb 2023

32. Zeng, D., Zhang, H., Xiang, L., Wang, J., Ji, G.: User-oriented paraphrase generation with keywords controlled network. IEEE Access **7**, 80542–80551 (2019)

33. van der Lee, C., Gatt, A., van Miltenburg, E., Krahmer, E.: Human evaluation of automatically generated text: current trends and best practice guidelines. Comput. Speech Lang. **67**, 101151 (2021)

Machine Learning Methods for Crop Yield Prediction

Vijayatai Hukare⬤, Vidya Kumbhar(✉) ⬤, and Sahil K. Shah⬤

Symbiosis Institute of Geoinformatics, Symbiosis International (Deemed University),
Pune, India
kumbharvidya@gmail.com

Abstract. As the world's population continues to increase, the demand for food production is on the rise, which requires attention to the challenges in agriculture. Fortunately, traditional agricultural practices can be managed more effectively with state-of-the-art techniques. These activities include irrigation management, crop yield improvement, pest and weed control, and fertilizer recommendations, among others. Improving crop yield is a significant aspect of agriculture management, and the parameters of soil, water, and climate play a vital role in achieving this goal. The application of information technology-based decisions and future predictions related to agricultural management can help farmers improve crop productivity by managing these complex systems. The Internet of Things (IoT), Data Mining, Cloud Computing, and Machine Learning (ML) are among the state-of-the-art techniques playing an essential role in agriculture. In this study, various ML models, such as Random Forest Regression, Gradient Boosting Regression, Adaboost Regression, and Decision Tree Regression, were employed to predict sugarcane yield. The study established a correlation between sugarcane yield and diverse climate and soil parameters. Based on a comparative analysis of the ML algorithms, the Gradient Boost Regression algorithm provided greater accuracy compared to other models. The study concludes that early prediction of sugarcane yield can help farmers increase their crop yield and subsequently improve their socioeconomic status.

Keywords: Agriculture · Crop Yield · Prediction · Sugarcane · Machine Learning

1 Introduction

Agriculture has existed since ancient times. The primary source of supplies for meeting people's basic requirements is thought to be agriculture. The Rigvedic elucidates various agricultural activities, such as ploughing, irrigation, and cultivation of fruits and vegetables. Not only Rigveda but also there are traces of cultivation during the Indus Valley civilization thriving period [1].

Crop yield improvement is one of the major activities of agriculture management. Soil parameters like Nitrogen (N), Phosphorus (P), Potassium (K), moisture of the soil, soil texture, etc., and climate parameters such as rainfall, humidity, temperature, etc. play

M. K. Saini et al. (Eds.): ICA 2023, CCIS 1866, pp. 195–209, 2023.
https://doi.org/10.1007/978-3-031-43605-5_15

an important role for crop yield improvement [2]. Accurate yield estimation, improved productivity and improved decision-making are essential in agriculture. Indian farmers are facing several problems such as unpredictable climate, limited irrigation facilities, and quality of water and soil [3]. In this scenario to improve productivity, the role of the latest information technologies, like the Internet of Things (IoT), Data Mining, cloud computing, Machine Learning, etc. have been vital over the years. In crop yield prediction, machine learning plays an essential role [4–9].

Ananthara et al. [10] have used the parameters like soil type, soil and water pH value, rainfall, and humidity and achieved an accuracy of 90%. Everingham et al. [11] developed a yield prediction model for sugarcane with the help of a random forest (RF) ML algorithm. The parameters used in this research are maximum and minimum temperature, rainfall, radiation, Southern Oscillation Index. The performance of the model is based on RMSE and R Squared. Natarajan et al.[12] developed a crop prediction model with the help of a machine learning technique using a hybrid approach. Researchers have used Genetic Algorithm (GA) and Driven Nonlinear Hebbian Learning (DDNHL) for the hybrid approach. The model's performance was based on RMSE, MAE, and Classification Accuracy (CA). Kumar et al. [13] developed a crop yield prediction model using a ML technique and the parameters used in this research are rainfall, soil parameter like (PH, N, P, K, OC, Zn, Fe, Cu, Mn, & S). Charoen-Ung & Mittrapiyanuruk developed the sugarcane yield prediction model in the year 2018 and 2019 using different parameters such as Soil type, plot area, groove width, water type and archived an accuracy of 71.83%. And Plot characteristics, Plot cultivation scheme, and rain volume and archived an accuracy of 71.88% [14, 15]. Kale & Patil proposed a prediction model using Artificial Neural Networks (ANN) algorithms. The parameters used in this research are Cultivation area, crops, State, District, Season & The performance of the model is based on RMSE, MAE, MSE and achieved an accuracy of 90% [16]. Khaki & Wang used deep neural networks to predict crop yield in the United States and Canada based on genotype and environmental data. Authors have used Lasso, shallow neural networks (SNN), and regression tree (RT) for the comparison and the performance of the model- based RMSE and achieved 12% accuracy [17]. Medar et al. [18] proposed the sugarcane yield prediction model with the help of a long-term time series and support vector Machine (SVM) ML algorithm. The parameters used in this research are temperature and soil temperature of the soil, evapotranspiration, humidity, moisture of the soil, sunshine duration, dew temperature, and precipitation. Saranya & Sathappan built an ensemble MME-DNN yield prediction model with the use of soil, weather and climate data [19]. Rale et al. [20] developed a prediction model for crop yield production by using machine-learning techniques and comparing the model performance of different linear and non-linear regression models using 5-fold cross-validation. Kang et al. [21] studied the effect of climatic and environmental variables on maize yield prediction. Researchers have used Long-short term memory (LSTM), Lasso, Support Vector Regressor, XGBoost, Convolutional Neural Network (CNN), Random Forest for the comparison and the performance of the model-based on RMSE, MAE, R, MAPE. Prasad et al. [22] proposed a cotton crop yield prediction model using random forest (RF) machine learning algorithms with the help of climate data. The performance of the model is based on the coefficient of determination (R^2) and achieved an accuracy of

0.83. Kanimozhi & Akila developed crop yield prediction model using Artificial Neural Networks (ANN) algorithm. The parameters used in the research are Location, Humidity, rainfall, Mintemp, Maxtemp. Wind_speed, Mean_temp. The performance of the model is based on RMSE & archived an accuracy of 80% [23]. Agarwal & Tarar developed a hybrid prediction model using the machine learning algorithm such as Support Vector Machine (SVM) and Deep learning algorithm such as Long-short term memory (LSTM). Researchers have used Random forest (RF) for the comparison. The performance of the model was based on precision, recall and achieved an accuracy of 97% [24]. Ansarifar et al. [25] presented a comparative analysis of linear regression, stepwise regression, Lasso regression, Ridge Random forest, XGBoost, Neural network, and Interaction regression for the prediction of the crop yield using different datasets of the crops. To calculate the performance of the model based on relative root mean square error (RRMSE). Interaction regression reached the minimum errors across the produced crop yield models. Dash et al. [26] proposed the yield prediction(YP) model using SVM and a decision tree ML algorithm. The parameters used in this research are rainfall, humidity, temperature, sunlight, and soil pH. Summary of literature review shown in the Table 1.

Table 1. Machine Learning Algorithm for Crop Yield Prediction

Research work	Algorithms used	Feature used in research	Crop used	Evelution Parameters	Accuracy
[10]	C&R tree, beehive clustering	Soil type, soil pH, water pH, Rainfall and Humidity	Rice, Sugarcane	Not Available	90%
[11]	Random forest (RF)	Temperature (maximum and minimum), Rainfall, Radiation,	Sugarcane	RMSE, R^2	95.45%
[12]	Hybrid (DDNHL-DA)	soil, pH, N, P, K, OC,Zn, Fe, Cu, Mn, S, Soil moisture Humidity, Rainfall Temperature, EC	Sugarcane	RMSE, MAE,, Classification Accuracy (CA)	94.70%
[13]	K-Nearest Neighbor, Support Vector Machine, Least Squared Support Vector Machine	Rainfall, Depth of soil, pH, N, P, K, OC, Zn, Fe, Cu, Mn, S	Sugarcane	MSE	90%

(continued)

Table 1. (*continued*)

Research work	Algorithms used	Feature used in research	Crop used	Evelution Parameters	Accuracy
[14]	Random Forest, Gradient boosting tree	Type of Soil, groove width, plot area, water type	Sugarcane	Not Available	71.83%
[15]	Random Forest	Characteristics of Plot, Plot cultivation scheme, Rain volume	Sugarcane	Not Available	71.88%
[18]	Support vector regression (SVR)	temperature, Soil temperature, soil moisture, humidity, dew temperature, sunshine duration, precipitation, evapotranspiration	Sugarcane	Not Available	83.49%
[24]	Hybrid prediction model using (SVM, RNN and LSTM)	Temperature, Rainfall, pH value, humidity, area	Wheat, Rice, Maize, Millets, Sugarcane	performance metrics (accuracy, precision, recall)	97%
[26]	SVM and decision tree	Rainfall, Humidity, Temperature, Sunlight, Soil pH	rice, wheat, and sugarcane	Not Available	92%

The comprehensive literature review confirms that Random Forest (RF), and decision tree regression machine learning algorithms are the most widely algorithms used for the sugarcane crop yield prediction. The review also confirms that the major parameters used for prediction are Rainfall, humidity, temperature, sunlight, soil (pH, N, P, K, OC, Zn, Fe, Cu, Mn, & S), type of the soil.

Main contributions of this work are:

- Analysing a correlation between sugarcane crop yield and various soil and climatic parameters.
- Prediction of sugarcane yield by utilizing regression machine-learning algorithms.

Our work specifically focuses on the western regions of Satara district, Maharashtra, India, which is known for its diverse agro-climatic areas. Further we examined three regions within this area, including plain regions (PR), drought-prone regions (DPR), and sub-mountain regions (SMR), all of which are known for cultivating sugarcane as a major crop.

2 Materials and Methods

Figure 1 shows the methodology followed for this study. The data used for the study are climate, soil and crop yield data. After the data collection, the zone wise data pre-processing was done & the Pearson Correlation Coefficient (PCC) was calculated. After that developed the prediction model for sugarcane yield prediction using ML algorithm & Model performance is measured using evaluation metrics & Generated the result.

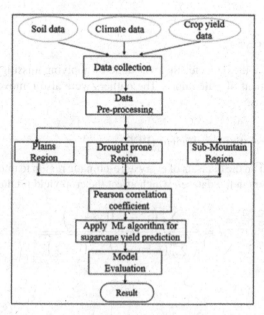

Fig. 1. Methodology

2.1 Data Collection

The climate, soil and crop yield data are used in study from the year 2000 to 2015. Table 2 shows the source description of the data collected for the study area for all the three regions.

Soil moisture (soil_m), precipitation (ppt), vapor pressure (vap), actual evapotranspiration (aet), maximum temperature (tmax), and minimum temperature (tmin) these parameters were used in the current study for the sugarcane yield prediction (YP).

Table 2. Source of the Dataset

Sr. no	Dataset	Sources
1	Soil data	Soil Survey and Soil Testing Laboratory, Satara
		www.soilhealth.dac.gov.in
2	Climate Data	Indian Metrological Department
3	Crop Yield data	District Agriculture Department of Kolhapur

2.2 Data Pre-processing

After data collection, the data cleaning was done by applying missing value processing methods like treatment of null values. The outliers were also removed to get the data ready for the analysis.

2.3 Pearson Correlation Coefficient (PCC)

PCC was calculated on the datasets of every agro-climatic region to understand the effect of distinct parameters in the dataset, which affect the crop yield prediction (Eq. 1) [28].

$$r_{xy} = \frac{\sum_i^n \left(x_i - x'\right)\left(y_i - y'\right)}{\sqrt{\sum_i^n \left(x_i - x'\right)^2}\sqrt{\sum_i^n \left(y - y'\right)^2}} \tag{1}$$

where,

n = Sample size.
x_x, y_i = Individual sample points indexed with i.

2.4 Crop Yield Prediction Using Machine Learning (ML) Algorithm

As the predicted value of yield will be Numerical, the category of algorithms selected will be regression type. The Gradient Boosting regression, Random forest, Decisiontrees Regression, Adaboost Regression machine learning (ML) algorithm is applied in current study for sugarcane yield prediction (YP).

Gradient Boosting Regression: Gradient boosting regression is a ML algorithm that allows for the optimization of any differentiable loss functions by building an additive model in a forward stage-wise manner. The negative gradient of the supplied loss function is used to fit a regression tree at each stage [29].

Algorithm:

GraB (T is a terminal region, $\propto$ is a learing rate)

Steps:

1. Initialize the first model:

$$f_0(x) = \text{argmin} \sum_{i=1}^{n} L(y, \gamma)$$
$$\gamma$$

2. Calculate the pseudo residuals for m^{th} model:

For m = 1, 2......, M
  i.   For i= 1,2.......... N calculate

$$Rim = -\left[\frac{dL(y, f(x_i))}{df(x_i)}\right]_{f=f_{m-1}}$$

  ii. Fit a regression tree to R_{im} & find terminal region:
       $T_{Im} =$ J= 1,2,.................Jm
  iii. For J =1, 2,.......Jm calculate
       $$\gamma_{Jm} = \text{argmin} \sum_{i=1}^{n} L(y, f_{m-1}(x_i) + \gamma)$$
       $$\gamma$$
  iv. Update the model for all data point
       $$f_m(x) = f_{m-1}(x) + \alpha \gamma_{Jm} I(x \sum T_{Jm})$$
3. After applying m^{th} model final prediction is:

$$F(x) = f_m(x)$$

Random Forest Regression: Random forest is an ensemble machine-learning algorithm that creates a regression tree using the subset of feature and bootstrap samples [8, 29].

The algorithm is as follows: we choose a bootstrap sample from Ts, where Ts(i) stands for the ith bootstrap, for each tree in the forest. Then, using a modified decision-tree learning algorithm, we learn a decision tree. The procedure is changed as follows: at each node of the tree, we randomly select a subset of the features $f \subseteq F$, where F is the set of features. Then, rather than splitting on F, the node splits on the best feature in f. After that, the node splits based on f's best feature rather than F's. F is significantly smaller in practice than f. Sometimes the most computationally expensive part of decision tree learning is choosing which feature to split. We significantly speed up the learning of the tree by reducing the collection of characteristics.

Algorithm:

RF (Training set $T_S = (x_1, y_1),(x_2, y_2) \ldots ,(x_n, y_n)$, features F, and B number of decision trees)

Steps:

1.   function RandomForest(T_S , F)
2. $L_t \leftarrow \emptyset$
3.   for $i \in 1, \ldots ,$ B do
4. $T_S^{(i)} \leftarrow$ bootstrap sample from T_S
5. $l_i \leftarrow$ RandomizedTreeLearn($T_S^{(i)}$, F)
6. $L_t \leftarrow L_t \cup \{l_i\}$
7.   end for
             i.   return L_t
8.   end function
9.   function RandomizedTreeLearn(T_S , F)
10.   For every node:
11.   f is a narrow subset of F
12.   Split on the best aspect of f
13.   return (learned tree)
14.   end function

Decision Tree Regression: Decision trees are supervised machine learning algorithms that are based on information. They have a tree structure comparable to a flow diagram, with each internal node indicating an attribute analysis, the branch depicting the test outcome, and each terminal node defining the class name [8, 29].

The decision tree has two stages:

i.   Build a Decision Tree
ii. Pruning Decision Tree

The training dataset is split recursively using the best criterion in the first stage until all or almost all of the data in each partition have the same class label. Consecutive branches are minimised during the pruning stage to build the tree in a way that effectively generalises the model. To improve specific criteria in the decision tree, pruning often entails traversing the decision tree from the bottom up or from the top down while deleting the noisy and outlier nodes.

Algorithm:

DT (S is a sample, F is a features)
Steps:
1. If End_Point (S, F) = true then
   i. L = CNode()
   ii. lLabel = divide(s)
   iii. return L
2. root = CNode ()
3. root(test_condition) = discover the best split (S, F)
4. R = {r | r a probable result for the root(test_condition)}
5. For every value r $\in$ R:
   i. S_r = {s | root (test_condition) (s) = r and s $\in$ S }
   ii. Child = TreeGrowth(S_r ,F)
   iii. Add a child as the root's descent and label the   edge {root →child} as r
6. return root

Adaboost Regression: The AdaBoost regression is a meta-estimator that starts by fitting a regression on the original dataset, then fits further copies of the regression on the same dataset, but with the weights of instances changed based on the current prediction's error [29]. By adjusting the weight of the training set, the AdaBoost algorithm is started. The training set (x_1, y_1), (x_2, y_2)..., (x_n, y_n) where each x_i corresponds to instance space X and each label y_i is in the label set Y, which is identical to the set of $\{-1, +1\}$. It designates W_m as the weight for the training example in round m as $W_m(i)$. At the beginning $(W_m(i) = 1/N, i = 1,..., N)$, the same weight will be set.

Algorithm:

Adaboost (Training set $T_S = x_i$ (i = 1, 2... n), labels $y_i \in Y$, number of Iteration =m)
Steps:
1. Sample $T_S = (x_1, y_1),(x_2, y_2) \ldots ,(x_n, y_n)$ where $x_i \in X$ and y_i {-1,+1}
2. Sample weight $W_1(i)$= 1/N where i = 1,.....N
3. for m = 1 to M
4. By using W_m to train weak learner
5. Weak hypothesis h_m :X → {-1, +1} with its training error:$\in m = \sum_i^n h_m(x) \neq y_i$
6. Update W_m:$W_{m+1}(i) = \frac{W_m \, e^{\alpha m} \, I_{mf_m}(x_m)}{C_m}$
7. Next m that, m +1
8. Output of hypothesis: $H_{(x)} = Sign \sum_{m=1}^M \alpha_m h_m(x)$

2.5 Model Evaluation

The efficiency of a machine-learning model is determined by comparing it to various performance criteria or utilizing various evaluation methods. Performance of above models is assessed using following evaluation metrics:

Adjusted R^2: The R-squared adjusted for the number of predictors in the model is known as adjusted R-squared (Eq. 2) [28].

$$\text{Adjusted R}^2 = \left\{ 1 - \left[\frac{(1 - R^2)(n - 1)}{(n - k - 1)} \right] \right\} \tag{2}$$

where,

R^2 = Sample R-square
k = Number of predictors variables
n = Total sample size

Mean Absolute Error (MAE): Mean Absolute Error is the arithmetical mean of the absolute difference between the actual and predicted observations (Eq. 3) [4, 28].

$$\text{MAE} = 1/n \sum_{j=1}^{n} |y_j - y_j'| \tag{3}$$

where,

$i = 1, 2 \dots n$ observations
y_j = actual observation
y_j' = predicted observation

Root Mean Squared Error (RMSE): Root Mean Square Error is the measure of how well a regression line fits the data points (Eq. 4) [28].

$$\text{RMSE} = \sqrt{\sum_{j=1}^{n} \frac{\left(y_j - y_j'\right)^2}{n}} \tag{4}$$

where,

y_j = actual observation
y_j' = predicted observation

3 Results and Discussion

3.1 Pearson Correlation Coefficient (PCC)

Pearson correlation coefficient of soil & climate parameters with a crop yield of agro-climatic regions is depicted in Table 3. The result shows that climate parameters such as tmax and tmin of this parameter are positively correlated in DPR and negatively correlated in PR and SMR. It indicates that climate parameters i.e. tmax & tmin are more effective parameters for DPR than PR and SMR. Soil parameters like soil moisture (soil_m) and climate parameters like precipitation (ppt) are positively correlated in PR, DPR and SMR. It indicates that these two parameters are more effective parameters of all three region. In the PR and DPR climate parameters such as vapor pressure (vap) and evapotranspiration (aet) are positively correlated and negatively correlated SMR. It indicates that vapor pressure and evapotranspiration more effective parameters of PR and DPR as compared to SMR.

Table 3. Correlation between Soil & Climate Parameters with Crop Yield for the Study Area

Agro-climatic Regions	Parameters					
	P1(tmax)	P2(tmin)	P3(soil_m)	P4(ppt)	P5(vap)	P6(aet)
PR	−0.0945	−0.0744	0.16468082	0.0847	0.1835	0.01406
DPR	0.27706	0.29546	0.23695853	0.3129	0.42572	0.36806
SMR	−0.2846	−0.2801	0.24612039	0.3012	-0.2728	-0.3654

3.2 Relationship Between Soil and Climate Parameters with Crop Yield for the Study Area

The correlation of the parameters for the Plains, Drought Prone, and Sub-Mountain Agro-climatic regions. For plains regions (PR), it is observed that soil parameters like soil moisture (soil_m) and sugarcane yield is positively correlated. The climate parameters such as precipitation (ppt), vapor pressure (vap), and actual evapotranspiration (aet) are positively correlated with sugarcane yield, were as maximum temperature (tmax), and Minimum temperature (tmin) are negatively correlated. The negative correlation of temperature with crop yield indicates that as the temperature reduces the water holding capacity of the soil is retained and which is very useful for sugarcane yield prediction. For drought prone regions (DPR) the soil parameters like soil moisture(soil_m) and climate parameters such as precipitation (ppt), vapor pressure (vap), actual evapotranspiration (aet), maximum temperature (tmax), and minimum temperature (tmin) are equally important for the sugarcane crop production in the DPR. Due to water scarcity in the drought prone regions the smallest change in these parameters affect the crop yield.In correlation between sugarcane crop yield with soil and climate parameters of the sub mountain regions (SMR). Here it has been observed that soil parameters like soil moisture (soil_m) and climate parameters such as precipitation (ppt) are positively correlated with sugarcane yield. The other climate parameters such as vapor pressure (vap), and actual evapotranspiration (aet), maximum temperature (tmax), and minimum temperature (tmin) are negatively correlated with sugarcane yield. It indicates that soil moisture and precipitation are effective parameters for the sugarcane crop production as compared to the other climate parameter such as vap, aet, tmin & tmax in the sub mountain regions.

3.3 Evaluation of Machine Learning Models

Table 4 shows the performance of the model of sugarcane yield prediction for Plain, Drought Prone, and Sub Mountain agro-climatic regions. Gradient Boosting regression showed accuracy scores of 85.33% for Plains, 69.62% for Drought Prone, and 61.68% for Sub-Mountains agro-climatic regions respectively and the lowest accuracy scores were of Random Forest regression having 38.19% for Sub Mountains, 74.68% for Plains, 62.47% for Drought Prone, and 38.19% for Sub Mountains. It is observed that the accuracy score of the Plains regions is highest as compared to the other regions such as Drought Prone and Sub-Mountain regions.

Table 4. Adjusted R^2, MAE, RMSE value for Plain, Drought Prone, and Sub Mountain Regions

Algorithms	Matrix	Agro-climatic regions		
		Plains	Drought Prone	Sub-Mountain
Random Forest	Adjusted R^2	0.74	0.62	0.38
	MAE	9.09	15.84	12
	RMSE	10.12	22.15	14.79
Gradient boosting Regression	Adjusted R^2	0.85	0.69	0.61
	MAE	8.58	15.31	11.38
	RMSE	9.75	22.34	13.42
Adaboost Regression	Adjusted R^2	0.75	0.61	0.43
	MAE	8.56	15.35	12.96
	RMSE	9.73	22.86	15.32
Decision Tree Regression	Adjusted R^2	0.89	0.64	0.49
	MAE	7.23	15.84	13.28
	RMSE	8.3	24.31	15.2

Figure 2 depicts Actual vs Predicted crop yield. It is observed that predicted yield values are influenced positively by sample number and the trend between actual and predicted values is directly related. It can be inferred from the figure that actual values show a significant drop for sample numbers ranging from 6 to 9. Actual and predicted crop yields converge almost at the same point for sample number 11.

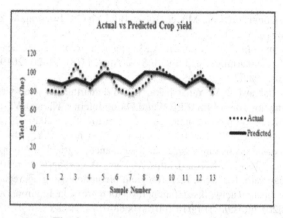

Fig. 2. Actual vs Predicted crop yield

4 Conclusion

This study highlights the significance of considering the diverse agro-climatic regions in predicting crop productivity. The Gradient Boosting algorithm was found to be more effective than other machine learning algorithms in predicting sugarcane yield in the plain, drought-prone, and sub-mountain regions, with accuracy scores of 85.33%, 69.62%, and 61.68%, respectively. Moreover, the study identified crucial soil and climate parameters, such as soil moisture, precipitation, vapor pressure, actual evapotranspiration, and temperature, that influence sugarcane yield. The results imply that farmers can use the prediction model developed in this study to enhance sugarcane productivity. However, the study has limitations, including its focus on only the western parts of Satara district in Maharashtra and limited consideration of soil and climate parameters. Future research should address these limitations by incorporating these parameters to develop a more comprehensive sugarcane yield prediction model.

References

1. Kale, S.S., Patil, P.S.: A machine learning approach to predict crop yield and success rate. In: 2019 IEEE Pune Sect. Int. Conf. PuneCon 2019, pp. 1–5 (2019). doi: https://doi.org/10.1109/PuneCon46936.2019.9105741
2. van Klompenburg, T., Kassahun, A., et al.: Crop yield prediction using machine learning: a systematic literature review. Comput. Electron. Agric. **177**, 105709 (2020). https://doi.org/10.1016/j.compag.2020.105709
3. Deepa, N., N, S. K., Srinivasan, K., Chang, C.Y., Bashir, A.K.: An Efficient ensemble VTOPES multi-criteria decision-making model for sustainable sugarcane farms. Sustainability **11**(16) (2019). https://doi.org/10.3390/su11164288
4. Gonzalez-Sanchez, A., Frausto-Solis, J., Ojeda-Bustamante, W.: Predictive ability of machine learning methods for massive crop yield prediction. Spanish J. Agric. Res. **12**(2), 313–328 (2014). https://doi.org/10.5424/sjar/2014122-4439
5. Groenendyk, D., Thorp, K., Ferré, T., Crow, W., Hunsaker, D.: A k-means clustering approach to assess wheat yield prediction uncertainty with a HYDRUS-1D coupled crop model. In:

Proc. - 7th Int. Congr. Environ. Model. Softw. Bold Visions Environ. Model, iEMSs 2014, vol. 3, pp. 1326–1333 (2014)

6. Medar, R.A., Rajpurohit, V.S., Ambekar, A.M.: Sugarcane crop yield forecasting model using supervised machine learning. Int. J. Intell. Syst. Appl. **11**(8), 11–20 (2019). https://doi.org/10.5815/ijisa.2019.08.02

7. Haque, F.F., Abdelgawad, A., Yanambaka, V.P., Yelamarthi, K.: Crop yield analysis using machine learning algorithms. In: IEEE World Forum Internet Things, WF-IoT 2020 - Symp. Proc., pp. 31–32 (2020). https://doi.org/10.1109/WF-IoT48130.2020.9221459

8. Elavarasan, D., Durai Raj Vincent, P.M., Srinivasan, K., Chang, C.Y.: A hybrid CFS filter and RF-RFE wrapper-based feature extraction for enhanced agricultural crop yield prediction modeling. Agric. **10**(9), 1–27 (2020). https://doi.org/10.3390/agriculture10090400

9. Khosla, E., Dharavath, R., Priya, R.: Crop yield prediction using aggregated rainfall-based modular artificial neural networks and support vector regression. Environ. Dev. Sustain. **22**(6), 5687–5708 (2020). https://doi.org/10.1007/s10668-019-00445-x

10. Ananthara, M.G., Arunkumar, T., Hemavathy, R.: CRY-An improved crop yield prediction model using bee hive clustering approach for agricultural data sets. In: Proceedings of the 2013 International Conference on Pattern Recognition, Informatics and Mobile Engineering, PRIME 2013, pp. 473–478 (2013). https://doi.org/10.1109/ICPRIME.2013.6496717

11. Everingham, Y., Sexton, J., Skocaj, D., Inman-Bamber, G.: Accurate prediction of sugarcane yield using a random forest algorithm. Agron. Sustain. Dev. **36**(2) (2016). https://doi.org/10.1007/s13593-016-0364-z

12. Natarajan, R., Subramanian, J., Papageorgiou, E.I.: Hybrid learning of fuzzy cognitive maps for sugarcane yield classification. Comput. Electron. Agric. **127**, 147–157 (2016). https://doi.org/10.1016/j.compag.2016.05.016

13. Dubey, S.K., Sharma, D.: Assessment of climate change impact on yield of major crops in the Banas River Basin, India. Sci. Total Environ. **635**, 10–19 (2018). https://doi.org/10.1016/j.scitotenv.2018.03.343

14. Charoen-Ung, P., Mittrapiyanuruk, P.: Sugarcane yield grade prediction using random forest and gradient boosting tree techniques. In: 2018 15th International Joint Conference on Computer Science and Software Engineering (JCSSE), July 2018, pp. 1–6. https://doi.org/10.1109/JCSSE.2018.8457391

15. Charoen-Ung, P., Mittrapiyanuruk, P.: Sugarcane yield grade prediction using random forest with forward feature selection and hyper-parameter tuning. Adv. Intell. Syst. Comput. **769**, 33–42 (2019). https://doi.org/10.1007/978-3-319-93692-5_4

16. Kale, S.S., Patil, P.S.: A Machine Learning Approach to Predict Crop Yield and Success Rate. 2019 IEEE Pune Section International Conference. PuneCon **2019**, 1–5 (2019). https://doi.org/10.1109/PuneCon46936.2019.9105741

17. Khaki, S., Wang, L.: Crop yield prediction using deep neural networks. Front. Plant Sci. **10**, 1 (2019). https://doi.org/10.3389/fpls.2019.00621

18. Medar, R., Rajpurohit, V.S., Shweta, S.: Crop yield prediction using machine learning techniques. In: 2019 IEEE 5th International Conference for Convergence in Technology (I2CT), pp. 1–5, March 2019. https://doi.org/10.1109/I2CT45611.2019.9033611

19. Saranya, M., Sathappan, S.: Multi-model ensemble with deep neural network based crop yield prediction. Int. J. Adv. Sci. Technol. **28**(17), 411–419 (2019)

20. Rale, N., Solanki, R., Bein, D., Andro-Vasko, J., Bein, W.: Prediction of crop cultivation. In: 2019 IEEE 9th Annual Computing and Communication Workshop and Conference, CCWC 2019, pp. 227–232, March 2019. https://doi.org/10.1109/CCWC.2019.8666445

21. Kang, M.O., Zhu, X., Ye, Z., Hain, C., Anderson, M.: Comparative assessment of environmental variables and machine learning algorithms for maize yield prediction in the US Midwest. Environ. Res. Lett. **15**(6) (2020). https://doi.org/10.1088/1748-9326/ab7df9

22. Prasad, N.R., Patel, N.R., Danodia, A.: Crop yield prediction in cotton for regional level using random forest approach. Spat. Inf. Res. (2020). https://doi.org/10.1007/s41324-020-00346-6
23. Kanimozhi, E., Akila, D.: An empirical study on neuroevolutional algorithm based on machine learning for crop yield prediction. Lecture Notes in Networks and Systems **118**, 109–116 (2020). https://doi.org/10.1007/978-981-15-3284-9_13
24. Agarwal, S., Tarar, S.: A hybrid approach for crop yield prediction using machine learning and deep learning algorithms. J. Phys. Conf. Ser. **1714**(1) (2021). https://doi.org/10.1088/1742-6596/1714/1/012012
25. Ansarifar, J., Wang, L., Archontoulis, S.V.: An interaction regression model for crop yield prediction. Sci. Rep. **11**(1), 1–14 (2021). https://doi.org/10.1038/s41598-021-97221-7
26. Dash, R., Dash, D.K., Biswal, G.C.: Classification of crop based on macronutrients and weather data using machine learning techniques. Results Eng. **9**, 100203 (2021). https://doi.org/10.1016/j.rineng.2021.100203
27. Comprehensive District Agriculture Plan (2016–2017) ,satara District. (2016)
28. Gupta, S.C., Kapoor, V.K.: Fundamentals of mathematical statistics. New Delhi, India: Sultan Chand & Sons (P) (2020)
29. Pradhan, M., Dinesh Kumar, U.: Machine Learning using Python. Wiley (2019)

Real-Time Plant Disease Detection: A Comparative Study

Yogendra Singh[1]([✉]), Swati Shukla[2], Nishant Mohan[1],
Sumesh Eratt Parameswaran[2], and Gaurav Trivedi[1]

[1] Indian Institute of Technology, Guwahati, India
yogendrasingh@iitg.ac.in
[2] VIT-AP University, Vellore, AP, India
https://www.iitg.ac.in/, https://vitap.ac.in/

Abstract. Plant diseases account for over 30% of production loss in India. Early detection of these diseases is crucial to maintaining yield. However, manual surveillance is laborious, costly, time-consuming and requires domain knowledge. Computer vision offers a non-destructive and efficient solution to disease detection, with classical machine learning and deep learning (DL) algorithms. DL methods offer several advantages, especially in scenarios where there is a large amount of data to process. With automatic feature extraction, these techniques can efficiently analyze multi-dimensional inputs, reducing the time and effort required for processing. Consequently, their usage has gained significant popularity in identifying and diagnosing diseases in plants. In this paper, we conduct a comprehensive comparative study of 14 cutting-edge object detection algorithms, including default and modified versions of YOLOv7 and YOLOv8. Our study focuses on their performance in real-time plant disease detection. The study involved several stages, including pre-processing, fine-tuning using pre-trained weights and validation on two publicly available datasets, namely PlantDoc and Plants Final, comprising real-life images of plant leaves. In particular, the study compared the performance of default YOLO models with YOLO models that used default architecture after freezing the backbone weights during theAs per Springer style, both city and country names must be present in the affiliations. Accordingly, we have inserted the country name in the affiliation. Please check and confirm if the inserted country name is correct. If not, please provide us with the correct country name fine-tuning process. The results show that the modified YOLO models achieved comparable mean Average Precision (mAP) values to the default models on the Plants Final dataset while reducing training time and GPU load by 9–25% and 19–47%, respectively, depending on the size of the backbone in different models.

Keywords: computer-vision · deep learning · object detection · transfer learning · YOLO

Supported by TIC and E&ICT Academy, at IIT Guwahati.

1   Introduction

Being the main source of earning in rural areas, agriculture plays an important role in the development of developing countries like India. However, crop losses due to diseases, pests and dynamic weather conditions significantly affecting the contribution made by the agricultural sector. The losses in the agricultural sector can have a significant impact on agro-based countries like India. The cost of commonly consumed vegetables like onions, potatoes and tomatoes has significantly risen from their typical prices, possibly attributed to crop losses. Accurate and timely diagnosis of plant diseases is essential to prevent the wastage of resources and promote effective management.

The traditional method of plant disease diagnosis is limited by its labor-intensive and time-consuming nature. This leads to a high risk of yield losses in areas without access to technical advice. Laboratory-based tests have complex methods and time requirements, making them less practical. While some diseases may have no visible symptoms or delayed effects, accurate diagnosis can be challenging due to variations in symptoms, making it necessary for trained plant scientists to carry out sight examinations [1].

To address these challenges, precision agriculture practices have been increasingly adopted to achieve a sustained increase in efficiency and yields. Precision agriculture involves the utilisation of advanced technology sensors and analysis tools to improve crop yields and aid in management decision-making. It is a recent concept that has been implemented globally with the aim of boosting production, reducing labor requirements and optimising fertiliser and irrigation management.

Plants often exhibit visible signs on leaves, trunks, flowers, or fruits, making visual examination enough to predict the disease. As a result, non-invasive methods have gained attention, with a focus on developing automated, fast and accurate mechanisms for disease detection.

Technologies such as Computer Vision, Machine Learning and Deep Learning which are sub-fields of Artificial Intelligence, have shown significant results in different sectors like healthcare, entertainment, defence, agriculture etc., to enhance decision-making, automate tasks and improve overall efficiency. With the help of computer vision, machines can perceive and interpret the world around them, enabling them to perform tasks that were once exclusive to humans.

Image processing techniques are one of the most popular methods being used for plant disease detection. High-quality cameras with sensitive sensors have been developed, which can capture images in various formats such as visible light, spectral, thermal and fluorescence imaging. These images are then processed using a variety of image processing approaches and used to train and test Machine Learning algorithms. However, earlier work using classical machine learning techniques faced limitations due to smaller datasets and hand-crafted feature extraction methods, leading to limited performance and scope for detecting various types of crops and diseases [2].

With recent advances in technology such as GPUs (Graphics Processing Units), increased availability of data, new algorithms etc., deep learning has overcome limitations faced by machine learning, enabling the development of more sophisticated and accurate automated systems. Deep learning algorithms can extract meaningful features automatically from images and identify patterns that are not easily detectable by human inspection. Therefore, the use of deep learning in conjunction with high-quality imaging systems and image processing techniques is a promising approach for improving plant disease detection accuracy and reducing yield losses.

In the past decade, a number of convolutional neural network architectures have been developed for classifying images into different classes. Some of them are LeNet, AlexNet, VGG, GoogLenet and ResNet. DenseNet, in particular, has been noted to achieve comparable accuracy to ResNet while utilising fewer parameters. These architectures have improved prediction performance by incorporating non-convolutional layers, residual learning and batch normalisation techniques. Moreover, they have reduced computational complexity by employing smaller convolution filters as compared to previous CNN architectures [3].

Object detection architectures have evolved significantly over the years, with both two-stage and one-stage models being used for this task. Two-stage models use convolution-based region proposal networks to filter regions of interest before passing them through the region of interest pooling network. However, this approach has a quantisation issue that affects region of interest prediction. To address this, a masked model based on a convolution network was developed with a ResNet backbone that enhances semantic segmentation and small object detection. It also uses pixel-to-pixel alignment instead of ROI pooling, which considers all computational values of the features [4].

One-stage models like YOLO and SSD enable real-time detection of multiple classes from input images without prior predetermination. YOLOv3, for instance, uses cross-entropy functions and a feature pyramid network-like system to make feature extraction more vigorous. It is faster than single shot detector due to its darknet backbone. RetinaNet is another model that uses a feature pyramid network as a backbone classifier to improve cross-entropy performance and decrease missed classified cases during training. Several deep learning architectures have been developed based on standard models to enhance ROI determination and disease identification. This comparison was made by L. Tan et al. [5].

The development of these models has significantly contributed to the advancement of computer vision and object detection techniques. But the emergence of YOLOv7 [6] and YOLOv8 [7] has expanded the possibilities in the area of Object Detection.

This study compares different versions of YOLOv7 and YOLOv8 for the viability of object detection to detect and classify diseases. For this comparison, two datasets are used, namely PlantDoc [8] and Plants Final [9]. These datasets have real-life plants images which can relate to real-time disease detection tasks. The paper is structured into six sections, beginning with the introduction in Sect. 1. Section 2 provides a comprehensive review of relevant literature. Section 3 details the dataset used in our study. The methodology employed in the experiments and evaluation metrics is outlined in Sect. 4. Section 5 reports the experimental results, while Sect. 6 presents the conclusions and outlines potential areas for future research.

2   Literature Review

The advancements in AI have brought significant breakthroughs in deep learning and computer vision technologies, addressing complex issues such as voice and face recognition, NLP, medical image processing and translation, Autonomous vehicles etc. Convolutional neural networks have demonstrated their effectiveness in several industries, including automotive, healthcare, finance etc. and are now being utilised in agriculture for the automated detection of crop diseases, which provides a viable alternative to conventional methods. Recent research has introduced various models and applications for crop disease identification and diagnosis, which will be discussed in this section.

In recent years, crop disease identification and diagnosis using deep learning models have shown immense potential, with several studies exploring different approaches. For instance, VGG-16 was used in a deep convolutional neural network framework by B. S. Anami et al. [10] to classify various stresses in paddy crops. A. Fuentes et al. [11] used ResNet for classification combined with single-stage and two-stage Object detection algorithms in tomato plant disease detection. D. Oppenheim et al. [12] developed an improved VGG network-based algorithm for accurate and quick identification and classification of spots on potato crops.

Several studies demonstrate the potential of deep learning models for crop disease identification and diagnosis and highlight the importance of continued research to develop more accurate and efficient methods for disease detection in agriculture. Researchers have explored methods to improve disease recognition and classification accuracy. J. Chen et al. [13] proposed a DenseNet-based method for detecting rice plant diseases. A region growing algorithm based on CNN architecture to identify cucumber disease spots in greenhouse setups was developed by J. Ma et al. [14]. VGGNet was used by C. R. Rahman et al. [15] in a disease recognition algorithm combined with InceptionV3 with reduced parameters and improved classification accuracy for rice plants. X. Fan et al. [16] used images with complex backgrounds and proposed an improved CNN architecture based model to identify few common corn diseases. A. I. Khan et al. [17] used the Xception model for leaf classification in a two-stage apple disease detection system combined with the Faster-RCNN model for disease detection.

A. Abbas et al. [18] utilised a deep learning approach that incorporates Conditional Generative Adversarial Network (GAN) to produce synthetic images of tomato leaves for disease detection. C. Liu et al. [19] has presented a novel approach to identify instances of cucumber leaf diseases that are present in complex backgrounds. This method involves using an EFDet, which stands for Efficient Feature Detection. First, the efficient backbone network extracts the features from the input and then the feature fusion module combines these features at different levels to provide a complete representation. Finally, the predictor uses this comprehensive representation to detect the presence of cucumber leaf diseases. On the other hand, M. P. Mathew et al. [20] developed a YOLOv5-based disease detection model that can identify bacterial spot disease in bell pepper plants from the symptoms seen on the leaves. In addition, R. Barosa et al. [21]

proposed a framework for an aquaponics system that employs image processing and decision tree methodology for detecting diseases in eggplant, chilli, citrus and mandarin and then generating a report that is sent to the owner via a mobile application if a disease is detected.

A novel system for detecting plant diseases using threshold segmentation and random forest classification was developed by S. Kailasam et al. [22]. Their approach achieved an impressive recognition accuracy of over 97% and a true optimism rate of over 99%, along with a PSNR, SSIM and MSE of over 59, 0.99 and 0.008. A. M. Roy et al. [23] proposed a high-performance object detection framework that addressed several common issues in plant pest and disease monitoring using the modified YOLOv4 algorithm. The proposed model achieved an accuracy of over 90%, an F1 score of over 93% and an mAP value of over 96% at a detection rate of 70 FPS by incorporating DenseNet, SPP and modified PANet.

Recent works on the PlantDoc dataset have shown the immense potential for object detection in detecting plant diseases. A. Shill et al. [24] proposed M_YOLOv4 model based on YOLOv4, which achieves an mAP value of 55.45%. It was also seen in the above study that YOLO versions overpowered faster RCNN and were better in Object detection with higher fps. D. Wang et al. [25] proposed the TL-SE-ResNeXt-101 model, which employs transfer learning and residual networks, achieves an mAP of 47.37% when the input image dimensions are 224 * 224. S. Vaidya et al. [26] used an augmented PlantDoc dataset achieving an mAP value of 71%. No research has been published yet on the Plants Final dataset, which is comparatively a new dataset.

The literature review has shown that deep learning techniques have been extensively used by researchers for detecting and classifying plant or crop diseases. The analysis also revealed that most disease detection systems are developed for open-air farms, with only a few systems designed for modern farming systems like aquaponics or hydroponics. Furthermore, most models are developed to identify multiple diseases of a single crop. Detecting diseases in crops is challenging due to several reasons. One of them is the similarity in foliage among different crops, which can impact the detection system's performance. Additionally, the visual symptoms of different diseases may appear similar due to variations in light illumination during imaging. Moreover, there is a lack of large-scale open-source datasets available for training deep learning models. While PlantVillage and PlantDoc have been instrumental in advancing the field of plant disease detection and developing more effective detection models, recently published datasets such as Plants Final have not received as much attention. As a result, there has been limited research on this dataset and its potential for improving plant disease detection. However, it is possible that Plants Final, along with other newly developed datasets, may provide valuable insights and opportunities for further enhancing the accuracy and efficiency of plant disease detection models. Therefore, it is important for researchers to explore the potential of all available datasets in order to make the most significant advancements in this field.

3   Dataset Analysis

In this paper, we utilised two publicly available datasets for Plant Disease Detection: the PlantDoc dataset and the Plants Final dataset. Both datasets are extensively described in this section to provide a comprehensive understanding of the data sources used in our research. PlantDoc dataset comprises a collection of images of diseased and healthy plant leaves from 13 different plant species captured in diverse environments and under varying conditions. On the other hand, the Plants Final dataset is a large-scale dataset consisting of images of diseased and healthy plant leaves from 13 plant species. It combines images from the PlantDoc dataset and several other sources. These datasets are valuable resources for the development and evaluation of machine learning algorithms for plant disease detection and have been widely used in the research community. By utilising these datasets, we aimed to contribute to the development of accurate and efficient models for plant disease detection, ultimately improving crop yield and food security.

3.1   PlantDoc Dataset

The PlantDoc dataset was published by researchers from IIT Gandhinagar in November 2021. One of the authors of the PlantDoc dataset made it publicly available. There are 2598 images included in this dataset that pertain to 13 different plant species and may exhibit up to 17 different disease classifications. These images were obtained through web scraping and have been annotated [8]. It includes both healthy and diseased categories, suitable for image classification and object detection. A total of 8,851 labels are available. The distribution of the PlantDoc dataset is shown in Fig. 1.

The PlantDoc dataset was carefully curated to address real-world issues. The raw images, referred to as uncropped images, were used to assess the deep learning methods performance. The dataset includes a variety of leaf image samples, such as leaf image records from the natural environment, leaf images with white backgrounds, other objects in the image and composites of leaf images.

For the model, a modified version of the PlantDoc Dataset was used with some corrections and adjustments made to the original Dataset by Roboflow. To improve the accuracy of the annotations, more than 28 annotations were rectified. Some bounding boxes were adjusted as they were marginally out of the frame and had to be cropped to align with the image's edge. In some cases, the bounding boxes mistakenly surrounded zero pixels and had to be removed. These corrections included 25 annotations in the training set and three in the validation set.

Sample labelled images from the PlantDoc dataset are shown in Fig. 2.

3.2   Plants Final Dataset

Plants Final Dataset is a publicly available dataset provided by Roboflow on Mar 5, 2023. In this paper, we have used "640 and Augmentation" version, which has

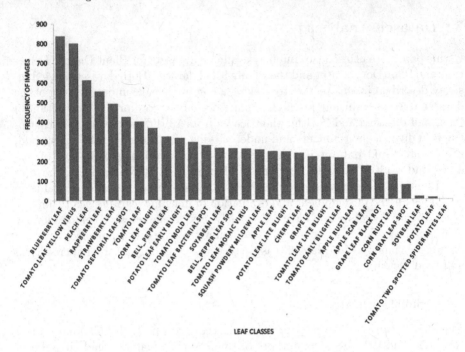

Fig. 1. Class Distribution in PlantDoc

augmented images of size 640 * 640 which is the default for both YOLOv7 and YOLOv8. This dataset contains 5705 images with the split of approximately 5000 training, 367 validation and 364 test set. Images in the dataset are preprocessed using auto orientation and resizing to 640 * 640. Horizontally flipped images are used here to make training more robust. Class Distribution for the Plants Final dataset is shown in Fig. 3. Sample labelled images from the Plants Final dataset are shown in Fig. 4.

4   Methodology and Evaluation Metrics

4.1   Methodology

This section provides an overview of the methodology used in the study and the modifications made to the default YOLOv7 and YOLOv8 models. The Plant-Doc dataset was initially subjected to augmentation by applying horizontal and vertical flips with a probability of 0.5 for each image during an epoch. This approach helped to increase the dataset's size and diversity, thereby improving the robustness of the model to variations in the input data. The Plants Final dataset was already horizontally augmented, so a vertical flip with a probability of 0.5 was introduced to further increase the model's ability to predict images in any orientation.

Fig. 2. Sample Images from PlantDoc Dataset

After the augmentation, fine-tuning over pre-trained weights was carried out in two phases. In the first phase, default settings were used after hyperparameter evolution, which involved finding the best possible settings for the model's hyperparameters during a 300 epochs trial run. This phase helped to improve the model's overall performance by optimising the settings for each component of the model.

In the second phase, the weights of the architecture backbone were frozen, as it is responsible for extracting low-level features that are similar regardless of the object being detected. This approach helped to reduce the load on the GPU and shorten the training time while maintaining the model's performance. Overall, these modifications were primarily aimed at improving the efficiency and effectiveness of the model for plant disease detection which is a critical area

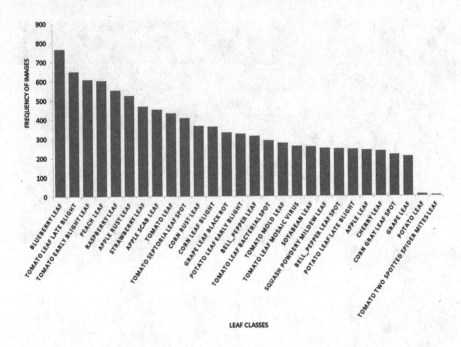

Fig. 3. Class Distribution in Plants Final

of research with significant implications for agricultural productivity and food security.

4.2   Evaluation Metrics

For evaluating a model's performance, we look at three metrics: Precision, Recall and Mean Average Precision (mAP). These metrics consider true positive (TP), false positive (FP), false negative (FN), average precision (AP) and intersection over union (IoU).

Precision measures how well the model can identify true positives out of all positive predictions.

$$Precision = \frac{TP}{TP + FP} \tag{1}$$

Recall measures how well the model can identify true positives out of all correct predictions.

$$Recall = \frac{TP}{TP + FN} \tag{2}$$

IoU calculates the overlap between the actual and predicted bounding box coordinates. The higher the overlap, the more accurate the bounding box.

$$IoU = \frac{Area\,of\,Intersection}{Area\,of\,Union} \tag{3}$$

The mAP is a widely used metric for object detection systems. It considers various factors like Precision, Recall, false positives, false negatives and IoU. We calculate the area under the Precision-Recall curve and average the precision for all classes at a specific IoU value. This provides a comprehensive assessment of the model's ability to detect objects in an image.

$$mAP = \frac{1}{n} \sum_{k=1}^{k=n} AP_k \tag{4}$$

5   Experimental Results

In this paper, fine-tuning and validation were conducted on a total of 28 YOLO models that were pre-trained on the COCO [27] image dataset. These models comprise 8 YOLOv8 models and 6 YOLOv7 models on each dataset, which include "modified" versions with frozen backbone weights from the pre-trained architecture. The training process was executed using a GPU (Quadro RTX 6000, 24220 MiB) with a batch size of 16, image size of 416*416 for the PlantDoc dataset and 640*640 for the Plants Final dataset.

In this study, the process of hyperparameter evolution was conducted for a total of 300 epochs, which is a relatively large number of iterations. Hyperparameter evolution is a process of finding the best possible values for the hyperparameters of a machine learning model. It involves running multiple iterations of the model with different combinations of hyperparameters and selecting the combination that results in the best performance. However, despite the extensive experimentation, only minor changes were observed in a few of the hyperparameters. Thus default set of hyperparameters was used. Nonetheless, the hyperparameter evolution process is crucial for optimising the performance of the model and is an important step in the development of any machine learning system.

Post hyperparameter evolution process fine-tuning was done and the Results are shown in Table 1.

Sample label images and corresponding prediction images of the best-performing model, i.e. YOLOv8 modified on Plants-Final dataset, is shown in Fig. 4 and Fig. 5, respectively.

When the plot of various validation losses vs the number of epochs was plotted it was noted that models with complex structure were getting over-fitted on Data, whereas modified versions were fluctuating in validation losses.

The modifications made to YOLOv7 and YOLOv8 were found to significantly reduce training time and GPU memory usage. Specifically, the modified versions were observed to reduce training time by 9% to 25% and GPU memory usage by 19% to 47%, depending on the size of the backbone architecture.

Table 1. Evaluated results on different models

Dataset	Model	Precision	Recall	mAP@0.5
PlantDoc	YOLOv8 nano	0.705	0.564	0.646
	YOLOv8 nano modified	0.676	0.466	0.572
	YOLOv8 medium	0.715	0.586	0.675
	YOLOv8 medium modified	0.653	0.536	0.612
	YOLOv8 large	0.691	0.604	0.677
	YOLOv8 large modified	0.711	0.529	0.623
	YOLOv8 XL (x6)	0.745	0.649	0.708
	YOLOv8 XL (x6) modified	0.671	0.588	0.639
	YOLOv7-tiny	0.568	0.629	0.618
	YOLOv7-tiny modified	0.448	0.587	0.559
	YOLOv7	0.665	0.618	0.635
	YOLOv7 modified	0.643	0.517	0.569
	YOLOv7-x	0.714	0.602	0.652
	YOLOv7-x modified	0.667	0.543	0.609
Plants Final	YOLOv8 nano	0.688	0.479	0.598
	YOLOv8 nano modified	0.721	0.512	**0.62**
	YOLOv8 medium	0.767	0.547	0.666
	YOLOv8 medium modified	0.729	0.552	0.656
	YOLOv8 large	0.747	0.57	0.672
	YOLOv8 large modified	0.766	0.568	**0.674**
	YOLOv8 XL (x6)	0.769	0.573	0.685
	YOLOv8 XL (x6) modified	0.768	0.597	**0.699**
	YOLOv7-tiny	0.699	0.594	0.628
	YOLOv7-tiny modified	0.642	0.571	0.583
	YOLOv7	0.724	0.639	0.67
	YOLOv7 modified	0.658	0.625	0.624
	YOLOv7-x	0.705	0.627	0.672
	YOLOv7-x modified	0.716	0.578	0.617

The evaluation of the modified versions of YOLOv7 and YOLOv8 on the PlantDoc dataset and the Plants Final dataset revealed some interesting observations. Table 1 shows that the modified YOLOv7 versions showed a reduction in mAP values ranging from 6.6% to 10.3% on the PlantDoc dataset and 6.8% to 8.1% on the Plants Final dataset. Similarly, the modified YOLOv8 versions showed a decrease in mAP values ranging from 7.9% to 11.5% on the PlantDoc dataset.

However, it is noteworthy that the modified YOLOv8 versions performed better than the default versions in most cases when trained on the Plants Final

Fig. 4. Sample labelled images in Plants-Final dataset

dataset. Examination of the two datasets revealed that the primary difference between them was the image resolution and the number of images. The Plants Final dataset contained images with a higher resolution and a larger number of images than the PlantDoc dataset.

While YOLOv8 performed better in terms of mAP than YOLOv7 on both datasets, the difference was only marginal. Based on these observations, it can be inferred that the modifications made to the YOLOv7 and YOLOv8 models should be used with larger and higher-resolution datasets to achieve optimal results. However, further investigation is needed to fully understand the potential benefits of these modifications and their applicability to other object detection models and datasets.

Fig. 5. Sample predicted images in Plants-Final dataset

6 Conclusions and Future Work

In conclusion, this study demonstrated that freezing backbone structures in object detection can be a viable approach to achieving comparable results while reducing training time and GPU memory usage. Moreover, the findings suggest that this method may be applied to other object detection tasks beyond plant datasets. Additionally, the study identified that complex models were susceptible to over-fitting and thus employed early stopping to address this issue. The results also indicated that increasing the dataset size, either by enhancing image resolution or increasing the number of images, can improve object detection performance. Looking forward, future work will concentrate on expanding the dataset by incorporating more real-life images and making logical modifications to object detection models. These efforts may lead to better performance in object detection tasks, thus advancing the field.

References

1. Talaviya, T., Shah, D., Patel, N., Yagnik, H., Shah, M.: Implementation of artificial intelligence in agriculture for optimisation of irrigation and application of pesticides and herbicides. Artif. Intell. Agric. **1**(4), 58–73 (2020)
2. Obsie, E.Y., Qu, H., Zhang, Y.J., Annis, S., Drummond, F.: Yolov5s-CA: an improved Yolov5 based on the attention mechanism for mummy berry disease detection. Agriculture **13**(1), 78 (2022)
3. Alzubaidi, L., et al.: Review of deep learning: concepts, CNN architectures, challenges, applications, future directions. J. Big Data **8**, 1–74 (2021)
4. Jiao, L., et al.: A survey of deep learning-based object detection. IEEE Access **5**(7), 128837–68 (2019)
5. Tan, L., Huangfu, T., Wu, L., Chen, W.: Comparison of RetinaNet, SSD and YOLO v3 for real-time pill identification. BMC Med. Inform. Decis. Mak. **21**, 1–11 (2021)
6. Wang, C.Y., Bochkovskiy, A., Liao, H.Y.: YOLOv7: trainable bag-of-freebies sets new state-of-the-art for real-time object detectors. arXiv preprint arXiv:2207.02696. 2022 Jul 6
7. Glenn, J.: Ultralytics YOLOv8 (2023). https://github.com/ultralytics/ultralytics
8. Singh, D., Jain, N., Jain, P., Kayal, P., Kumawat, S., Batra, N.: PlantDoc: a dataset for visual plant disease detection. In: Proceedings of the 7th ACM IKDD CoDS and 25th COMAD 2020 Jan 5, pp. 249–253
9. Roboflow. Plants Final Dataset (2023). https://universe.roboflow.com/plants-images/PlantsFinal
10. Anami, B.S., Malvade, N.N., Palaiah, S.: Deep learning approach for recognition and classification of yield affecting paddy crop stresses using field images. Artif. Intell. Agric. **1**(4), 12–20 (2020)
11. Fuentes, A., Yoon, S., Kim, S.C., Park, D.S.: A robust deep-learning-based detector for real-time tomato plant diseases and pests recognition. Sensors **17**(9), 2022 (2017)
12. Oppenheim, D., Shani, G.: Potato disease classification using convolution neural networks. Adv. Anim. Biosci. **8**(2), 244–9 (2017)
13. Chen, J., Zhang, D., Nanehkaran, Y.A., Li, D.: Detection of rice plant diseases based on deep transfer learning. J. Sci. Food Agric. **100**(7), 3246–3256 (2020)
14. Ma, J., Du, K., Zheng, F., Zhang, L., Gong, Z., Sun, Z.: A recognition method for cucumber diseases using leaf symptom images based on deep convolutional neural network. Comput. Electron. Agric. **1**(154), 18–24 (2018)
15. Rahman, C.R., et al.: Identification and recognition of rice diseases and pests using convolutional neural networks. Biosys. Eng. **1**(194), 112–20 (2020)
16. Fan, X., Zhou, J., Xu, Y., Yang, J.: Corn Diseases Recognition Method Based on Multi-feature Fusion and Improved Deep Belief Network
17. Khan, A.I., Quadri, S.M.K., Banday, S., Shah, J.L.: Deep diagnosis: a real-time apple leaf disease detection system based on deep learning. Comput. Electron. Agric. **198**, 107093 (2022)
18. Abbas, A., Jain, S., Gour, M., Vankudothu, S.: Tomato plant disease detection using transfer learning with C-GAN synthetic images. Comput. Electron. Agric. **187**, 106279 (2021)
19. Liu, C., Zhu, H., Guo, W., Han, X., Chen, C., Wu, H.: EFDet: an efficient detection method for cucumber disease under natural complex environments. Comput. Electron. Agric. **1**(189), 106378 (2021)

20. Mathew, M.P., Mahesh, T.Y.: Leaf-based disease detection in bell pepper plant using YOLO v5. SIViP **1**, 1–7 (2022)
21. Barosa, R., Hassen, S.I., Nagowah, L.: Smart aquaponics with disease detection. In: 2019 Conference on Next Generation Computing Applications (NextComp) 2019 Sep 19, pp. 1–6. IEEE
22. Kailasam, S., Achanta, S.D.M., Rao, Rama Koteswara, P., Vatambeti, R., Kayam, S.: An IoT-based agriculture maintenance using pervasive computing with machine learning technique. Int. J. Intell. Comput. Cybern. **15**(2), 184–197 (2022)
23. Roy, A.M., Bose, R., Bhaduri, J.: A fast accurate fine-grain object detection model based on YOLOv4 deep neural network. Neural Comput. Appl. **1**, 1–27 (2022)
24. Shill, A., Rahman, M.A.: Plant disease detection based on YOLOv3 and YOLOv4. In: 2021 International Conference on Automation, Control and Mechatronics for Industry 4.0 (ACMI) 2021 Jul 8, pp. 1–6. IEEE
25. Wang, D., Wang, J., Li, W., Guan, P.: T-CNN: trilinear convolutional neural networks model for visual detection of plant diseases. Comput. Electron. Agric. **1**(190), 106468 (2021)
26. Vaidya, S., Kavthekar, S., Joshi, A.: Leveraging YOLOv7 for plant disease detection. In: 2023 4th International Conference on Innovative Trends in Information Technology (ICITIIT) 2023 Feb 11, pp. 1–6. IEEE
27. Lin, T.-Y., et al.: Microsoft COCO: common objects in context. In: Fleet, D., Pajdla, T., Schiele, B., Tuytelaars, T. (eds.) ECCV 2014. LNCS, vol. 8693, pp. 740–755. Springer, Cham (2014). https://doi.org/10.1007/978-3-319-10602-1_48

Fruit Segregation Using Deep Learning

Archana Dehankar[✉] ⓘ, Hemanshu Dharmik ⓘ, Shravani Chambhare ⓘ,
Darshna Kohad ⓘ, Janvi Hingwe ⓘ, and Kaustubh Dhoke ⓘ

Department of Computer Technology, Priyadarshini College of Engineering, Nagpur, India
archana.dehankar@pcenagpur.edu.in

Abstract. In the agricultural sector or applications, manpower is needed for segregating or sorting different types of fruits. The improved sorting of fruits has an impact on the quality evaluation. Automation enhances and improves the standard and the efficiency of manufacturing goods. The main objective of the proposed system is to replace the manual inspection system to reduce the manpower required by using computer vision technology in the field of agriculture and fruit industry. An automatic fruit quality inspection system is used for sorting and grading of fruits- apple, banana and orange. This will help to speed up the process, improve accuracy and efficiency in less time. In recent research, there have been several algorithms developed by researchers to sort fruits using computer vision. The identification of class of the fruit relies on commonly used features, such as color, texture and morphological features. The proposed system acquires images from a camera placed in front of a laptop, and after processing the images, it will segregate them into specific types of fruits and will detect the quality of fruit identified. The fruits dataset is collected from the data library (Kaggle) to train the system. Information and images of fruits are in the dataset. FCN algorithm reads the dataset and learns the content/features of the fruits- apple, banana and orange. Fruits features will be used to identify the fruit type and quality of fruits. So, the segregation of fruits is done based on features such as shape, size and color of fruits. In this paper, we proposed a deep learning-based approach for fruit segregation using FCNs. We trained our FCN model on a large dataset of fruit images and achieved a high accuracy rate of 88.41% on the test dataset.

Keywords: Deep Learning · FCN · Computer Vision · Python · OpenCV · Feature Extraction · Google Colab · YOLO · Image Processing

1 Introduction

The use of image processing has been developing steadily in a variety of fields, including industrial, medical, real-time imaging, texture classification, object recognition, etc. Another rapidly expanding area of research is image processing and computer vision. Many plant diseases affect fruits and vegetables. As diseases can be observed on plant leaves and fruits, disease detection is crucial in agriculture. Fruit illnesses can be caused by pathogens, fungi, bacteria, microorganisms, viruses, and unfavorable environments.

M. K. Saini et al. (Eds.): ICA 2023, CCIS 1866, pp. 225–238, 2023.
https://doi.org/10.1007/978-3-031-43605-5_17

It is becoming more appealing to use image processing technology and computer vision software to identify fruit illness and fruit quality. Therefore, it is required to design a quick and affordable technique for inspecting the quality of fruit. There are numerous manual techniques for spotting fruit illnesses in their early stages. Seeing fruits with the naked eye is a classic but ineffective method of spotting fruit diseases. Digital techniques can make illness detection more efficient, accurate, and time efficient. To sort and grade different fruits, a fruit quality inspection system that operates automatically can be deployed. For precise fruit disease detection and identification, researchers have created numerous algorithms and various image processing approaches.

A collection of machine learning algorithms called "deep learning," which is based on the artificial neural network (ANN), is used to extract the needed information from a huge quantity of data using a mix of different methodologies. With the introduction of the back propagation method, one of the optimization methods, ANN attained its pinnacle, yet it reaches a technological limit. Hence, a nonlinear function led to the kernel approach (Support Vector Machine, Gaussian Process, etc.), which then led to machine learning. Many computational issues, initialization issues, and local minima issues are some of ANN's drawbacks. Pre-training employing unsupervised learning, computer development, parallel processing using GPGPU (General-Purpose Computing on Graphics Processing Units), and the growth of big data are the factors that could solve the challenges. Many issues are addressed by deep learning studies that are not addressed by traditional machine learning techniques. Several deep learning algorithms, including DNN (Deep Neural Network), CNN (Convolutional Neural Network), and RNN, are used in various fields (Recurrent Neural Network). Particularly in the areas of image processing such as image classification and image recognition, the CNN method performs well. As a result, numerous studies utilizing CNNs have been carried out and a variety of CNN models have been constructed. Models are developed in accordance with the properties of the data and fields, and transfer learning utilizing deep learning is also used by importing a pre-training model built in a healthy environment.

To extract information from data, the feature extraction procedure is crucial. The use of deep learning algorithms can produce more objective and superior features. One benefit of using CNN is that it can automatically extract features and evaluate data without the need for professional knowledge for the input data. This procedure evaluates the effectiveness of machine learning. The paper discusses the segregation of fruits according to their size, color and shape. It will then display the name and quality of each fruit after segregating the various fruits in an image.

2 Literature Survey

Researchers have used various deep learning techniques, such as Convolutional Neural Networks (CNN) [1], to recognize and classify fruits based on their shape, size, color, and texture. Anand Upadhyay, Sunny Singh, and Shona Kanojia developed a CNN-based system to segregate ripe and raw bananas with high accuracy [2]. Rucha Dandavate and Vineet Patodkar developed a CNN-based fruit classification model that used data augmentation techniques to improve classification accuracy [3].

Y. Liu, X. Xu, L. Zhang, and W. Liang proposed an improved CNN-based model for fruit recognition that achieved a recognition rate of 80.8% [4].

Anuja Bhargava, Atul Bansal & Vishal Goyal developed a machine learning-based detection and sorting system that achieved an accuracy rate of 84.9% for five different fruits and vegetables [5].

Machine learning algorithms have also been used to sort multiple fruits and vegetables based on their quality and maturity [6–8].

Aafreen Kazi & Siba Prasad Panda proposed a system for determining the freshness of fruits in the food industry using transfer learning [9].

Researchers have also explored the use of deep learning techniques for fruit segmentation and packaging. Anand Upadhyay, Sunny Singh, and Shona Kanojia developed a fruit segregation and packaging machine that used a CNN-based system to segregate and package fruits automatically [10].

Other researchers have explored the use of hyperspectral imaging and data augmentation techniques to improve fruit recognition accuracy. J Steinbrener, K Posch, and R Leitner proposed a hyperspectral fruit and vegetable classification system using CNNs [11].

In addition, some researchers have explored the use of deep learning techniques for fruit defect identification and maturity detection. D. Sahu and R. M. Potdar proposed an image analysis-based system for defect identification and maturity detection of mango fruits [12].

One of the studies on fruit detection and segmentation using deep learning was conducted by Liu et al. [13], where they proposed an efficient method for apple detection and segmentation in the context of harvesting robots. The proposed method used a deep learning model based on the Faster R-CNN architecture, achieving high detection and segmentation accuracy.

Similarly, Norhidayah Mohd Rozi et al. [14] proposed an artificial intelligence-based approach for fruit segmentation using a deep learning model based on the Mask R-CNN architecture. The proposed method achieved high accuracy in segmenting different types of fruits, including apples, bananas, and oranges.

Dhanapal and Jeyanthi [15] conducted a survey on fruit recognition using deep learning approaches, where they reviewed different deep learning models, including CNN, RNN, and LSTM. They discussed the advantages and limitations of each model and highlighted the challenges faced in the field of fruit recognition using deep learning.

Another survey was conducted by Ukwuoma et al. [16], where they reviewed recent advancements in fruit detection and classification using deep learning techniques. They discussed different deep learning models, including CNN, R-CNN, and YOLO, and compared their performance in fruit detection and classification tasks.

Nur-E-Aznin Mimma et al. [17] proposed a fruit classification and detection application using a deep learning model based on the CNN architecture. The proposed method achieved high accuracy in classifying and detecting different types of fruits, including apples, bananas, and oranges.

Shaikh et al. [18] proposed a deep learning-based approach for mango grading, where they used a CNN model to detect and classify mangoes based on their ripeness. The proposed method achieved high accuracy in mango grading, which can help in optimizing mango supply chains.

Liu et al. [19] proposed a novel method for apple grading using an improved Faster R-CNN model. The proposed method achieved high accuracy in grading apples based on their size and shape, which can help in automating apple grading processes.

Dandawate [20] proposed a deep learning-based approach for mango fruit detection and disease classification. The proposed method used a CNN model to detect and classify mango fruits based on their health status, which can help in early disease detection and prevention.

Zhang et al. [21] conducted a survey on recent advances in fruit detection and segmentation, where they reviewed different deep learning models, including CNN, R-CNN, and Mask R-CNN. They discussed the advantages and limitations of each model and highlighted the future directions of research in the field of fruit detection and segmentation.

Finally, Lin et al. [22] proposed a robust fruit detection and segmentation method for apple harvesting using deep learning. The proposed method used a deep learning model based on the Mask R-CNN architecture, achieving high accuracy in fruit detection and segmentation, which can help in automating apple harvesting processes.

3 Methodology

3.1 Data Collection

For deep learning or machine learning, a large amount of data in the form of images is always needed as a primary requirement. The first method for collecting data involves capturing multiple images of fruits in perfect lighting conditions to obtain better results. However, capturing vast numbers of images can be very timeconsuming and not feasible for us. The second method is to use pre-made datasets that are available on various platforms that share datasets for research purposes. One such platform is Kaggle, which we have chosen to obtain our dataset for the proposed work. From Kaggle, a dataset with 13.6k images of three types of fruits - apples, bananas, and oranges is obtained. The dataset contains 6 classes of images, namely FreshApple, FreshBanana, FreshOranges, RottenApple, RottenBanana, and RottenOranges.

3.2 Data Pre-processing

Data preprocessing is a critical step in any machine learning or deep learning as it helps prepare the dataset for the model training process. In this paper of fruit segregation using deep learning, a pre-made dataset is used which reduces our workload in data preprocessing as it removes irrelevant or blurry images or labels each image indicating the corresponding fruit type. Instead, it focuses on resizing the images to ensure that all the images have the same dimensions. To tackle the issue of overfitting, which is a common problem in deep learning algorithms, data augmentation techniques such as rotation, flipping, and zooming are utilized This technique increases the overall size of the dataset and helps the model to generalize better. The RGB images are converted to grayscale using OpenCV to allow the model to learn better. Furthermore, we normalized the pixel values of the images to ensure that the model trained efficiently and accurately. Data splitting and class balancing are not necessary as the Kaggle dataset used had already been separated into training and testing sets, and the classes have an equal number of images.

Overall, by performing these data preprocessing techniques, it is ensured that our model could learn effectively from the dataset and that it could generalize well to new data. Proper data preprocessing can significantly impact the model's performance and accuracy, and therefore it is essential to consider these techniques during the initial stages of any deep learning project. Results of data pre-processing are shown in Fig. 1.

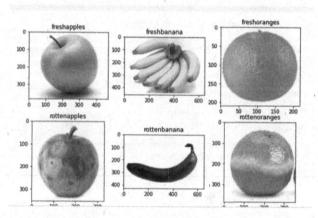

Fig. 1. (56 × 56) resize image.

3.3 Model Selection

The previous work in deep learning has mainly focused on using CNN (Convolutional Neural Network) and other similar algorithms. However, the proposed system on fruit segregation using deep learning, we aim to achieve better accuracy and therefore opted for FCN (Fully Convolutional Network), a type of neural network commonly used in computer vision tasks.

One of the key features of FCN is that it is a fully convolutional architecture that uses only convolutional layers for both feature extraction and classification. This is different from traditional CNNs, which often have a few fully connected layers at the end. The architecture of FCN is found to be particularly suitable as it can process images of arbitrary size, which is important for image segmentation tasks where the output mask needs to have the same size as the input image. The FCN architecture uses a combination of downsampling and upsampling layers to extract features from the input image and generate the output mask. This approach allows FCN to capture more spatial information from the input image, resulting in better segmentation accuracy.

Overall, FCN is a powerful algorithm that helped us achieve better accuracy. The model summary is shown in Fig. 2.

```
Model: "sequential_1"

Layer (type)                  Output Shape             Param #
=================================================================
rescaling_1 (Rescaling)       (None, 56, 56, 3)        0

conv2d_5 (Conv2D)             (None, 56, 56, 32)       2432

conv2d_6 (Conv2D)             (None, 56, 56, 32)       25632

max_pooling2d_4 (MaxPooling   (None, 28, 28, 32)       0
2D)

conv2d_7 (Conv2D)             (None, 28, 28, 32)       25632

max_pooling2d_5 (MaxPooling   (None, 14, 14, 32)       0
2D)

conv2d_8 (Conv2D)             (None, 14, 14, 32)       25632

max_pooling2d_6 (MaxPooling   (None, 7, 7, 32)         0
2D)

conv2d_9 (Conv2D)             (None, 7, 7, 32)         25632

max_pooling2d_7 (MaxPooling   (None, 3, 3, 32)         0
2D)

dropout_1 (Dropout)           (None, 3, 3, 32)         0

flatten_1 (Flatten)           (None, 288)              0

dense_1 (Dense)               (None, 6)                1734

=================================================================
Total params: 106,694
Trainable params: 106,694
Non-trainable params: 0
```

Fig. 2. Model Summary.

3.4   Hyperparameter Tuning

Hyperparameter tuning is the process of selecting the optimal values for the hyperparameters of a machine learning model. Hyperparameters are parameters that are not learned during model training, but instead must be set before training begins. Examples of hyperparameters in deep learning include learning rate, batch size, number of epochs, and network architecture.

The proposed system performs hyperparameter tuning to improve the accuracy of our model. We started by selecting a range of values for each hyperparameter and then ran multiple experiments using different combinations of hyperparameter values. For example, we experimented with different learning rates, batch sizes, and number of epochs to find the optimal values for our model. We also tried different network architectures, such as varying the number of convolutional layers, to see which configuration provided the best accuracy.

Based on the results of our experiments, the hyperparameters that provided the best accuracy are selected and then used those values to train our final model Fig. 3.

```
We trained model at 100 epochs.

Epoch 99/100
38/38 [==============================] - 1s 21ms/step - loss: 0.0028 -
accuracy: 0.9983 - val_loss: 0.6593 - val_accuracy: 0.8841
Epoch 100/100
38/38 [==============================] - 1s 21ms/step - loss: 0.0048 -
accuracy: 0.9975 - val_loss: 0.7431 - val_accuracy: 0.8841

Training Accuracy: 0.9975

Validation Accuracy : 0.8841
```

Fig. 3. Hyperparameters

3.5 Model Evaluation

The proposed system uses FCN (Fully Convolutional Network) as our deep learning algorithm for fruit segmentation. After training our FCN model on a large dataset of fruit images, its performance is evaluated using various performance metrics. One of the primary performance metrics used to evaluate the FCN model is pixel accuracy, which measures the percentage of pixels in the predicted mask that are correctly classified. We also used intersection over union (IoU) and mean intersection over union (mIoU) as performance metrics. IoU measures the overlap between the predicted mask and the ground truth mask, while mIoU is the average IoU over all classes of fruit.

In addition to these metrics, we used test images to visually inspect the model's segmentation output and ensured that it was accurately segmenting the fruit images. This visual inspection allowed us to identify any areas where the model was struggling and make adjustments to improve its performance. Overall, our evaluation of the FCN model using various performance metrics and visualization techniques allowed us to assess its accuracy and identify areas for improvement. By iteratively refining our model based on these evaluations, we were able to achieve a high level of accuracy in fruit segregation. Figure 4(a) shows the training and validation accuracy of the model. Figure 4(b) shows the accuracy of the model with test datasets.

3.6 Model Deployment

Model deployment involves making the trained model available for use by others. In the proposed system, the trained FCN model needs to be deployed so that it can be used to classify fruits in real-world scenarios. The Tkinter library is a popular Python library used to create graphical user interfaces. It provides a set of tools and widgets that can be used to develop interactive applications. The system uses Tkinter library to create an intuitive interface that enables users to easily input fruit images and obtain results. By developing a GUI for the trained FCN model using the Tkinter Python library it makes the fruit segregation process more accessible and user-friendly, allowing users with limited programming knowledge to make use of our model. Figure 5 shows the output of the model deployed using Tkinter. The GUI can be run on any device with Python and the required dependencies installed, making the model deployment process efficient and convenient for end-users.

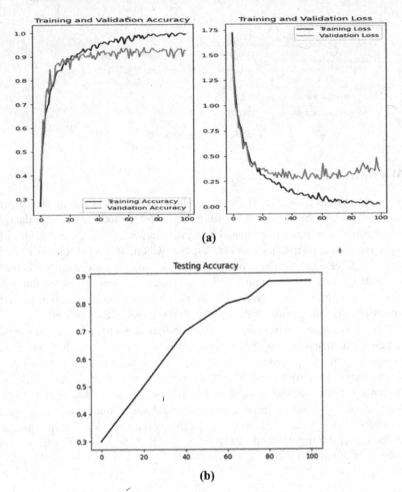

Fig. 4. (a) Training and Validation Accuracy (b). Testing Accuracy.

4 Technologies Used

4.1 Deep Learning

Deep learning is a subfield of machine learning that involves the use of artificial neural networks to model and solve complex problems. The term "deep" refers to the fact that these networks typically consist of multiple layers of interconnected nodes, with each layer responsible for extracting higher-level features from the input data.

Some of the most commonly used deep learning algorithms include convolutional neural networks (CNNs), recurrent neural networks (RNNs), and generative adversarial networks (GANs). These algorithms have been developed and refined over several decades, and they continue to be the subject of ongoing research and development in the field of deep learning.

Fig. 5. Model Deployment using Tkinter.

4.2 Python

Python is a high-level programming language used for various purposes such as web development, data analysis, artificial intelligence, scientific computing, and more. It is known for its simplicity, readability, and versatility. Python has a vast standard library and a huge community of developers constantly creating new libraries and tools. It supports multiple programming paradigms, including object-oriented, functional, and procedural programming. Python is open-source and cross-platform, meaning it can be used on different operating systems like Windows, macOS, and Linux.

4.3 OpenCV

A free and open-source software library for computer vision and machine learning is called OpenCV (Open Source Computer Vision Library). A wide range of computer vision and machine learning algorithms, both traditional and cutting-edge, are among the more than 2500 optimised algorithms in the library. These algorithms can be applied to recognise faces and objects, classify human activity in videos, track moving objects, extract 3D object models from stereo cameras, stitch images together to create high-resolution images of the entire scene, look for identical images in image databases, and remove red eyes from flash images.

4.4 YOLO

YOLO (You Only Look Once) is an object detection model that is widely used in computer vision applications. It was developed by Joseph Redmon, Santosh Divvala, Ross Girshick, and Ali Farhadi in 2016. YOLO is a single-stage detector that processes the entire image at once and outputs bounding boxes and class probabilities directly. It divides the input image into a grid of cells and for each cell predicts bounding boxes, class probabilities, and confidence scores. The confidence score reflects the model's confidence that a given bounding box contains an object.

4.5 Tkinter

Tkinter is a standard GUI (Graphical User Interface) library for Python programming language. It provides a simple way to create windows, dialogs, buttons, menus, etc. in a Python application. Tkinter is a builtin module in Python and does not require any external installation. With the help of Tkinter, you can create a GUIbased application that interacts with the user. It is a cross-platform GUI toolkit, which means that your code will work on all major operating systems like Windows, macOS, and Linux. Tkinter is widely used for developing desktop applications, games, scientific and engineering applications, and other applications.

5 Result

We evaluated our proposed system on a test dataset of fruit images. The test dataset consists of images that were not used for training the model. We achieved a high accuracy rate of 88.41% on the test dataset, which indicates that our model is highly effective in identifying and classifying fruits. Figure 6 shows the output of fruit segregation system.

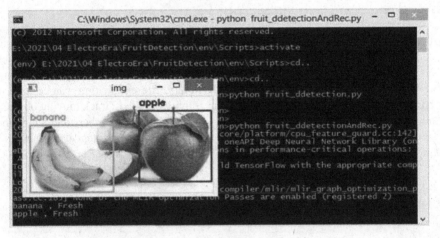

Fig. 6. Running system in python virtual environment.

In our work, we have implemented the YOLO object detection model for fruit detection, and for this purpose, we have utilized the cvlib library. Once the fruit is detected, we need to perform classification based on its quality. For building and training our model, we opted to use Google Colab. The reason behind choosing Google Colab is that it offers free GPU resources for 24 h, which greatly reduces the training time of our model. Therefore, we can train and fine-tune our model with large datasets in a shorter amount of time. Figure 7 shows the fruit detection in the system.

Fig. 7. Fruit Detection

The system uses an object detection model YOLO, to detect the presence of the apple and then uses a classification algorithm to classify it as rotten or not Fig. 8. Once the apple has been classified as rotten, it is then segregated from the other fruits to prevent it from being mixed with other fruits and potentially causing contamination or spoilage.

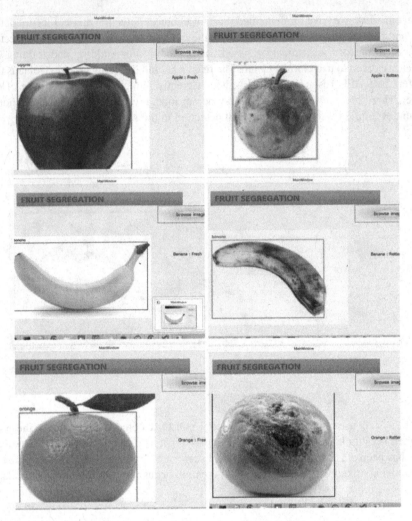

Fig. 8. Output images of fresh & rotten fruits.

6   Conclusion

The paper introduces an approach for fruit segregation using FCNs based on deep learning. To accomplish this, a large dataset of fruit images was utilized to train the FCN model, which resulted in an impressive accuracy rate of 88.41% on the test dataset. The important objective of this work is to recognize and categorize the fruits based on morphological features using significant feature extraction by Fully Convolutional Network (FCN) algorithm in deep learning. The fruit segregation system presented in this paper uses the enhanced fruit image detection techniques for extraction of morphological features of fresh or rotten fruits and feature selection to reduce irrelevant and redundant data from the feature set. This system may be adapted for accurate segregation of fruits based on color, shape, size thus replacing the human visual assessment.

References

1. Thangudu, R.K., Prakash, R.G., Stivaktakis, A.J.: Fruit recognition and classification using convolutional neural networks with feature selection. In: 2021 IEEE International Conference on Communication and Signal Processing (ICCSP) (2021)
2. Upadhyay, A., Singh, S., Kanojia, S.: Segregation of ripe and raw bananas using convolution neural network. In: International Conference on Machine Learning and Data Engineering Procedia Computer Science 218, pp. 461–468 (2023)
3. Dandavate, R., Patodkar, V.: CNN and data augmentation based fruit classification model. In: 2020 Fourth International Conference on I-SMAC (IoT in Social, Mobile, Analytics and Cloud) (I-SMAC). IEEE
4. Nordin, M.J., Xin, O.W., Aziz, N.: Food image recognition for price calculation using convolutional neural network. In: Proceedings of the 2019 3rd International 2019 – ACM
5. Ghodke, V., Pungaiah, S., Shamout, M., Sundarraj, A., Judder, M.I., Vijaprasath, S.: Machine learning for auto segregation of fruits classification based logistic support vector regression. In: 2022 2nd International Conference on Technology Advancements in Computational Sciences (ICTACS)
6. Aherwadi, N., Mittal, U.: Fruit quality identification using image processing, machine learning, and deep learning: a review. Adv. Appl. Math. Sci. 21(5), 2645–2660 (2022)
7. Don M. Africa, A., Tabalan, A.R.V., Tan, M.A.A.: Ripe fruit detection and classification using machine learning. Int. J. Emerging Trends Eng. Res. 8(5) (2020)
8. Bhargava, A., Bansal, A., Goyal, V.: Machine learning based detection and sorting of multiple vegetables and fruits. Food Anal. Methods 15, 228–242 (2022). Springer (Aug. 2021)
9. Kazi, A., Panda, S.P.: Determining the freshness of fruits in the food industry by image classification using transfer learning. Multimed. Tools Appl. 81, 7611–7624 (2022)
10. Upadhyay, A., Singh, S., Kanojia, S.: Design of fruit segregation and packaging machine. In: 2020 International Conference on Computational Performance Evaluation (ComPE) North-Eastern Hill University, Shillong, Meghalaya, India, 2–4 July, 2020
11. Steinbrener, J., Posch, K., Leitner, R: Hyperspectral fruit and vegetable classification using convolutional neural networks. Comput. Electron. (2019). Elsevier
12. Sahu, D., Potdar, R.M., Sahu, D., Potdar, R.M.: Defect identification and maturity detection of mango fruits using image analysis. Am. J. Artif. Intell. 1(1), 5–14 (2017)
13. Liu, K., Chen, X., Wang, Y.: An efficient fruit detection and segmentation method for apple harvesting robots based on deep learning. IEEE Trans. Ind. Inform. (2019)
14. Rozi, N.M., et al.: Development of fruits artificial intelligence segregation. Int. J. Nanoelectron. Mater. 14(Special Issue), 245–252 (2021)
15. Dhanapal, A., Jeyanthi, P.: A survey on fruit recognition using deep learning approaches. In: 2019 IEEE 2nd International Conference on Inventive Research in Computing Applications (ICIRCA) (2019)
16. Ukwuoma, C.C., Zhiguang, Q., Heyat, B.B., Ali, L., Almaspoor, Z., Monday, H.N.: Recent Advancements in Fruit Detection and Classification Using Deep Learning Techniques - Volume 2022 | Article ID 9210947
17. Mimma, N.-E-A., Ahmed, S., Rahman, T., Khan, R.: Fruits classification and detection application using deep learning. Hindawi Scientific Programming Volume 2022, Article ID 4194874

18. Shaikh, A., Banerjee, S., Samanta, K.: Mango grading using deep learning techniques. In: 2019 10th International Conference on Computing, Communication and Networking Technologies (ICCCNT), pp. 1–7. IEEE (2019)

19. Liu, S., Zhou, S., Huang, T., Wang, Y.: A novel method for apple grading based on improved Faster R-CNN. J. Food Eng. **290**, 110159 (2020)

20. Dandawate, N., Dandawate, Y.: Mango fruit detection and disease classification using deep learning. Procedia Comput. Sci. **156**, 18–27 (2019)

21. Zhang, H., et al.: Recent advances in fruit detection and segmentation: a survey. J. Food Eng. **271**, 109786 (2020)

22. Lin, T.-H., et al.: Robust fruit detection and segmentation for apple harvesting using deep learning. IEEE Robot. Autom. Lett. **5**(2), 2725–2732 (2020)

Investigation of the Bulk and Electronic Properties of Boron/Nitrogen/Indium Doped Armchair Graphene Nanoribbon for Sensing Plant VOC: A DFT Study

Jaskaran Singh Phull[1], Harmandar Kaur[1(✉)], Manjit Singh[1], Butta Singh[1], Himali Sarangal[1], Sukhdeep Kaur[2], Rupendeep Kaur[2], and Deep Kamal Kaur Randhawa[1]

[1] Department of Engineering and Technology, GNDU Regional Campus, Jalandhar, India
`{harmandar.ecejal,manjit.ecejal,rupendeep.elec}@gndu.ac.in`
[2] Department of Electronics Technology, GNDU, Amritsar, India
`sukhdeep.elec@gndu.ac.in`

Abstract. Studies have found that sulphur derivatives from dimethyl disulphide, an essential semiochemical aids in numerous significant plant growth functions such as enzyme activity, nitrogen metabolism, and synthesis of proteins. Various studies have confirmed that dimethyl disulphide is a vital fertiliser which aids in the growth and promotion of various plant species. Sensing the presence/prevailing sulphur content can essentially aid in regulating the inputs that are provided in field, thus contributing to the effective eco-friendly agrarian practices. The vitality and longevity of plant species are influenced by availability of this essential voc. In this paper, we investigate the adsorption behaviour of dimethyl disulphide molecule on boron doped graphene nanoribbon by using the density functional theory method. On a mono atom (boron/nitrogen/ indium) doped armchair graphene nanoribbon, a first-principles analysis based on density functional theory is conducted in order to recognise the presence of a volatile organic molecule named dimethyl disulphide which can assist in sensing plant growth. The investigation of the bulk, electronic and transport properties reveal the suitability of the nanoribbon for detection of this vital volatile organic compound. The resulting values are favourable because they indicate the existence of the adsorption process. The electronic characteristics of boron doped AGNR with dimethyl disulphide disclose p-type semiconducting behaviour, whereas the electronic properties of nitrogen doped AGNR with dimethyl disulphide reveal n-type semiconducting behaviour. Furthermore, the results of indium doping on AGNR using dimethyl disulphide show metallic characteristics.

Keywords: adsorption · DFT · armchair graphene nanoribbon · sensing · plant VOC

M. K. Saini et al. (Eds.): ICA 2023, CCIS 1866, pp. 239–251, 2023.
https://doi.org/10.1007/978-3-031-43605-5_18

1   Introduction

Recent investigations have revealed that semiochemicals are pivotal in intra- and inter-kingdom interactions in flora. These small odorous volatile organic compounds can modify the plant behaviour and can even promote/inhibit growth of the species in their vicinity. These primary plant communication mediators modulate the phenotypic plasticity and aid in vital ecological functions such as plant growth, abiotic stress tolerance management, and resilience to disease, defence mechanism against pathogens or herbivores.

VOCs act as chemical signalling agents that can travel far and interact with the neighbouring species acting as long-distance messengers. The interaction causes modulation of the behaviour of the interacting plant or microbe species. In plant kingdom, VOCs promote plant growth by acting as fumigant hence formulating plant defences against pathogens and herbivores. VOCs are generated by the strains of the plant growth promoting rhizobacteria. They are plant symbionts that improve crop weight, crop yield, abiotic stress tolerance and plant disease resistance, owing to which they can find application in the agriculture sector [1]. For instance, dimethyl disulphide (DMDS) is a VOC emitted by P. agglomerans species that plays an unequivocal role in the upkeep of the rhizosphere by enhancing sulphur nutrition [2]. Hence, from this pretext it is established that chemical signalling by the VOCs is utilised by plants to activate growth and for stress tolerance [3].

In this work, we aim at nano-scale sensing of DMDS, a chemical signalling agent that aids in the facile conduct of multiple aforementioned essential plant functions. The nanosensing can ascertain their role in interspecies (chemical signal aided) communication by aiding in construction of an effective olfactometer, thereby facilitating external supplementation in case of deficit of the said VOC [4].

Graphene is the first discovered two- dimensional nanomaterial that paved the path for the discovery of various other two-dimensional nanomaterials [5]. Graphene has unique properties and these can be engineered by various methods [6]. Graphene nanoribbons (GNR) are obtained by cutting monolayer graphene of finite length. [7]. GNRs possess unique properties such as high carrier mobility, semiconducting behaviour, flexible bandgap and mechanical robustness [8–10]. These properties make GNRs an important candidate in sensing application and reports in literature have explored the use of GNRs for the sensing of various gases [11–13]. The nanoribbons have been studied for the detection of VOCs as biomarkers [14–16]. Precisely, the use of graphene nanoribbons for sensing applications is highly favourable in the development of ultra-sensitive sensors with high packing density, better selectivity, higher sensitivity, faster recoverability, and less power consumption [17].

To the best of our knowledge, this work is a first in the semiochemical nanosensing of plant growth promoting DMDS molecule by doped armchair graphene nanoribbon by using density functional theory method. The doping is done by boron, nitrogen and indium in armchair graphene nanoribbon.

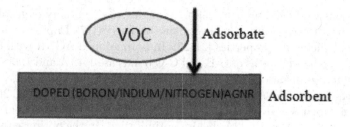

Fig. 1. Block diagram of adsorption of dimethyl disulphide VOC on doped AGNR

This work is vital for application in smart agriculture domain and can be adopted appropriately for enhanced farm practices by aiding in plant growth and defence mechanism. The rest of the paper is arranged as follows: Sect. 2 discusses the computational method used for the investigation of the bulk and electronic properties of the DMDS-doped AGNR complex; Sect. 3 covers the results and discussion for the nanosensing application which is followed by conclusion.

2   Computational Method

The investigation of the sensing of plant growth promoting DMDS by doped armchair graphene nanoribbon (AGNR) is conducted by using Virtual NanoLab Atomistix Toolkit [18]. In this investigation, the bulk and electronic properties of the considered complex are analysed after extensive calculations for establishing the use of the said nanoribbon in the sensing of DMDS. The bulk and electronic calculations are carried out by first principles using density functional theory (DFT) and non-equilibrium Green's function (NEGF). The generalised gradient approximation [19] of Perdew-Burke-Ernzerhof with double-zeta polarized basis set was used as the exchange-correlation basis set [20].

The doping in AGNR is performed using boron, nitrogen and indium single atom [21]. The calculations are performed on the geometrically relaxed structures with self-consistency. The density mesh cut off is set to 75 Hartree and the electron temperature is maintained at 300K. The values of force and stress tolerance are set as 0.05 eV/Å and 0.05 eV/Å3, respectively. The complex structures are allowed to be fully relaxed using geometry optimization before performing calculations of their electronic and transport properties. The k-point sampling of $1 \times 1 \times 12$ using Monkhorst-Pack grid is set for structural relaxation and $1 \times 1 \times 100$ for electronic transport calculations [22].

3   Results and Discussion

3.1   Geometric Structure Analysis

AGNR are built having six number of atoms along the width (w = 6). The edges are passivated by hydrogen atoms since the surfaces are less reactive than the edges. Various works in literature report that the sensing ability of graphene can be altered by selectively introducing dopants or defects [17]. Doping is included in GNR for enhancing its properties [21, 23]. In boron doped AGNR with dimethyl disulphide, the bond lengths of C-B and C-C bonds are 1.42 Å and 1.43 Å respectively, before optimization. In the optimised structure the bond lengths of C-B and C-C change to 1.51 Å and 1.41 Å respectively, as is shown in Fig. 2(a)–2(b). In nitrogen doped AGNR with dimethyl disulphide, the bond lengths of C-N and C-C bonds are 1.42Å and 1.42Å, before optimisation. In the optimised structure the bond lengths of C-N and C-C change to 1.36Å and 1.40Å respectively, as is shown in Fig. 2(c)–2(d). The bond length variations for indium doped AGNR with dimethyl disulphide before optimisation for C-In bonds and C-C bonds are 1.42Å and 1.42 Å and these values change after optimisation in the relaxed geometry and become 1.75 Å and 1.76 Å respectively as is shown in Fig. 2(e)–2(f). The relaxed structure of indium doped AGNR with dimethyl disulphide shows a bond formation as is evident in Fig. 2(f). The strong adsorption energy and reduction in the distance between molecule and indium doped AGNR upon relaxation suggests possible chemisorption in indium doped AGNR with dimethyl disulphide. This can form one-time use or disposable nanosensor device.

The changes in the bulk properties are a consequence of the influence of the adsorbate molecules on the adsorbent by the adsorption process. The resultant changes owing to the influence of the externally introduced molecule are an indicative of variations in the properties of the considered complex. Moreover, the effect of the gas molecule on the electronic properties of nanoribbon with bandgap is more than that on a nanoribbon with no bandgap. This property of the nanomaterials is imperative in various applications of nanosensing domain. The adsorption process upon the exposure to the guest molecule needs to be investigated in terms of the bulk and electronic properties that are computed by using the obtained relaxed structure shown in Fig. 1.

3.2   Adsorption of DMDS on Doped (Boron/Nitrogen/Indium) AGNR

In order to observe the adsorption of DMDS on different geometries of doped (boron/nitrogen/indium) AGNR, the first principles approach based on density function theory method is applied on the optimized complex structure. The adsorption on doped (boron/nitrogen/indium) AGNR after interaction with DMDS is investigated for bulk properties analysis. The adsorption energy calculation is performed by using the following formula:

$$E_{ad} = E_{D-AGNR+DMDS} - E_{D-AGNR} - E_{DMDS} \qquad (1)$$

Where ED-AGNR+DMDS, ED-AGNR and EDMDS denote the total energies of the doped (boron/nitrogen/indium) AGNR molecule complex, doped (boron/nitrogen/indium) AGNR, and the isolated DMDS VOC molecule, respectively. The comprehensive analysis of adsorption behaviour between DMDS molecule and boron/nitrogen doped AGNR requires the calculation of the

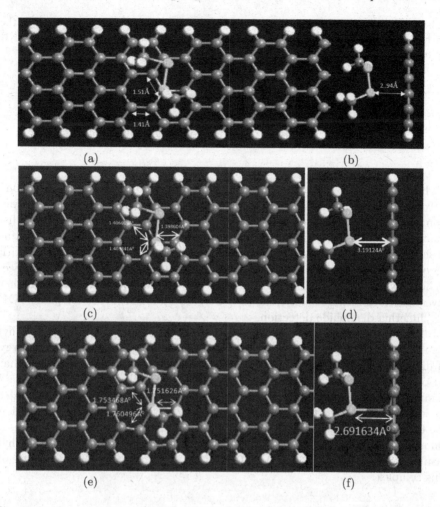

Fig. 2. Optimized AGNR with DMDS having (a) boron-doped top view (b) boron-doped side view (c) nitrogen-doped top view (d) nitrogen-doped side view (e) indium-doped top view (d) indium-doped side view

relaxed distance obtained after optimisation D(Å), energy gap (Eg) and adsorption energy (Ead). Table 1 presents the calculated values for all the mentioned parameters for doped (boron/nitrogen/indium) AGNR with DMDS molecule complex.

Table 1. The calculated band gap energy (Eg), the adsorption energy (Eads), the interaction distance of DMDS molecule with doped (boron/nitrogen/indium) AGNR for relaxed structure

Configuration	$E_g(eV)$	$E_{ad}(eV)$	D(Å)
Boron-AGNR with DMDS	0.864	−0.344	2.952
Nitrogen-AGNR with DMDS	0.823	−0.308	3.191
Indium-AGNR with DMDS	0.309	−0.962	2.691

In order to observe the adsorption of dimethyl disulphide on boron/nitrogen/indium doped AGNR, dimethyl disulphide gas molecule is placed away from doped AGNR before optimization. After optimization the calculated distance of dimethyl disulphide from boron doped AGNR has a value 2.95241 Å as is shown in Fig. 2(b). Also, the adsorption energy of dimethyl disulphide with boron doped AGNR is −0.34467 eV, indicating that adsorption is taking place. Hence, owing to the obtained favourable results this complex can be further investigated in terms of electronic properties to establish its use for plant VOC i.e. dimethyl disulphide detection.

Also, for nitrogen doped AGNR after optimization the distance of dimethyl disulphide has a value 3.19 Å as is shown in Fig. 2(d). The adsorption energy of dimethyl disulphide with nitrogen doped AGNR is −1.31 eV, indicating that adsorption process is taking place. Hence, in the light of the above results the said complex can be further investigated in terms of its electronic properties in order to establish its use for the detection of dimethyl disulphide. After optimization the calculated distance of dimethyl disulphide from indium doped AGNR has a value 2.69 Å as is shown in Fig. 2(f). Also, the adsorption energy of dimethyl disulphide with indium doped AGNR is −0.96 eV, indicating metallic nature of this complex.

3.3   Band Structure Analysis

Band gap is calculated to analyze the change in electronic properties of boron/nitrogen/indium doped AGNR before and after dimethyl disulphide adsorption. The band structure represents the energy of the available electronic states along a series of lines in a reciprocal space. The band structure presents the energy states of the complex formed in terms of band gaps or forbidden gaps. The band gap value and binding distance of boron doped AGNR, nitrogen doped AGNR and indium doped AGNR after dimethyl disulphide adsorption is shown in Table 1. The band gap value of boron doped AGNR is 0.864eV after dimethyl disulphide adsorption as shown in Fig. 3(a). The band gap value for nitrogen doped AGNR is 0.823eV after dimethyl disulphide adsorption, which is shown in Fig. 3(b) while the band structure for indium doped AGNR has a band gap value of 0.309 eV after dimethyl disulphide adsorption. Also, the fermi level shifts towards the valence band. The band structure analysis results for boron

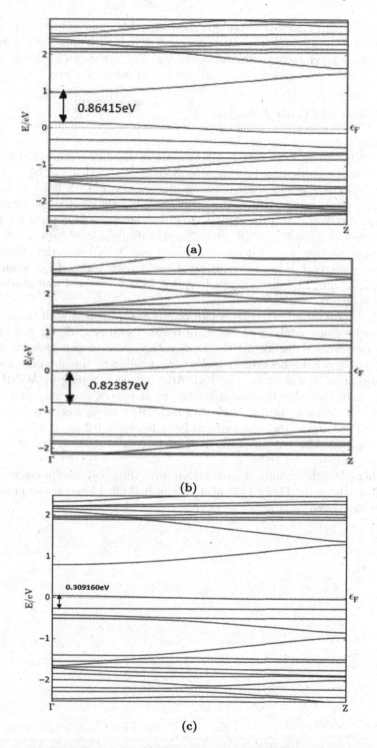

Fig. 3. Band structure of doped AGNR adsorbed dimethyl disulphide (a) boron doped AGNR (b) nitrogen doped AGNR (c) indium doped AGNR

doped AGNR with dimethyl disulphide show that this complex has p-type semi-conducting behaviour while the nitrogen doped AGNR with dimethyl disulphide complex shows n-type semiconductor behaviour. This is a useful finding that can be utilised in creating various electronic devices at nanoscale level

3.4   Density of States Analysis

To observe the interaction of dimethyl disulphide on boron/nitrogen/indium doped AGNR, the density of states (DOS) for this complex is studied. The DOS plot presents the energetically favourable available states that can be occupied. More precisely, it specifies the number of allowed electron (or hole) states per volume at a given energy. Hence, they indicate about the carrier concentrations and energy distributions of the carriers. Figure 4 shows the DOS plots for doped (boron/nitrogen/indium) AGNR with dimethyl disulphide adsorption. A comparison of boron doped AGNR and nitrogen doped AGNR dimethyl disulphide complex reveals that a greater number of peaks with higher DOS values are detected after dimethyl disulphide adsorption in valence band and conduction band, respectively.

Furthermore, the obtained DOS plot results are consistent with those of the band structure plot analysis of Fig. 3. In boron doped AGNR, the conduction band range shows reduction in the DOS values. Also, a gap appears in the conduction band lying between 0–1 eV corresponding to the absence of states in this region as is evident for Fig. 4(a). After nitrogen doping in AGNR with dimethyl disulphide, the density of states plot indicates large variation in the conduction band range of 0 to 2 eV and high DOS values between 1 to 2 eV. Also, a gap appears in the valence band lying between 0.2 to −1 eV as can be seen in Fig. 4(b). The gap corresponds to an absence of states in the mentioned range with zero value of DOS.

Furthermore, the indium doped AGNR with dimethyl disulphide has high DOS values about the Fermi level at 0 eV and half-filled band shows proximity to the valence band.

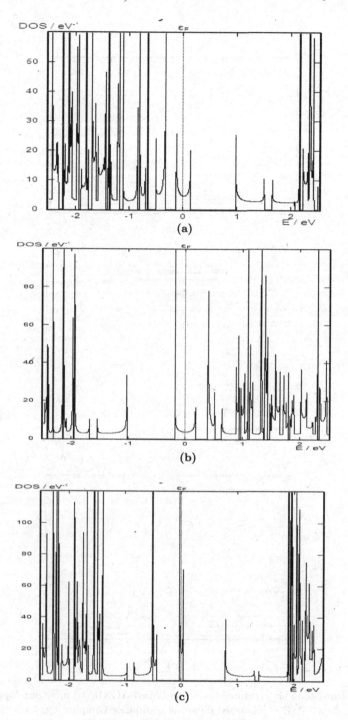

Fig. 4. DOS of (a) boron doped AGNR (b) nitrogen doped AGNR (c) indium doped AGNR with dimethyl disulphide complex

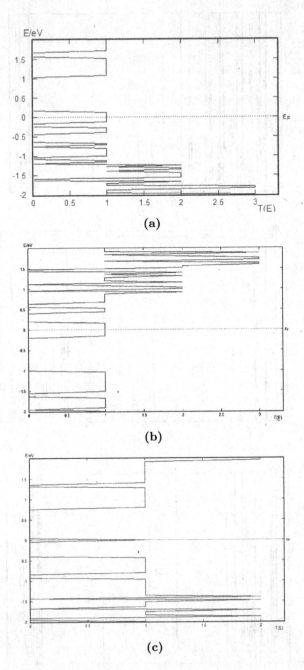

Fig. 5. Transmission spectrum of (a) boron doped AGNR (b) nitrogen doped AGNR (c) indium doped AGNR adsorbed dimethyl disulphide complex.

3.5   Transmission Spectrum Analysis

The transmission spectrum is a sum over the available modes in the band structure at each energy.

The transmission spectrum analysis for boron/nitrogen doped AGNR with dimethyl disulphide is presented in Fig. 5(a)–5(b). It is evident from the investigation that the transmission spectra plot corresponds to the resultant behaviour of the complex in both the obtained bandstructure analysis and the density of states analysis of Fig. 3 and Fig. 4 respectively.

The transmission coefficient has a zero value for the boron doped AGNR with dimethyl disulphide between 0.2 to 1 while for nitrogen doped AGNR with dimethyl disulphide the zero transmission window lies between 0 to −1 as is shown in Fig. 5(a) and Fig. 5(b), respectively. This analysis validates the previously obtained results and shows that boron doped AGNR with dimethyl disulphide portrays p-type behaviour while nitrogen doped AGNR with dimethyl disulphide presents n-type semiconducting behaviour. The transmission has high value around Fermi level as is evident form Fig. 5(c) for the indium doped AGNR with dimethyl disulphide.

4   Conclusion

The presence of dimethyl disulphide plays a pivotal role in the plant growth promotion and disease counter mechanism. However, the depleted levels or absence of this vital VOC can lead to inhibit plant growth, the same can be externally artificially supplemented for instance using fertilisers. The sensing of dimethyl disulphide using doped (boron/nitrogen/indium) AGNR is investigated by calculating the bulk and electronic properties upon relaxation of the complex. The bulk properties investigations include the calculation of the distance after relaxation, adsorption energy and energy bandgap. The values obtained are favourable as they reflect the occurrence of the adsorption process. The investigation of the electronic properties by the band structure analysis, density of states analysis and the transmission spectrum analysis for boron doped AGNR, nitrogen doped AGNR and indium doped AGNR with dimethyl disulphide is conducted. Upon investigation, the electronic properties reveal p-type semiconducting behaviour for boron doped AGNR with dimethyl disulphide while the electronic properties for nitrogen doped AGNR with dimethyl disulphide show n-type semiconducting behaviour. The results for indium doping on AGNR with dimethyl disulphide present metallic character. Thus, these investigations reveal the suitability of doped AGNR for detection of dimethyl disulphide. Furthermore, these findings can prove useful in the development of electronic devices in nanoscale domain.

References

1. Vasseur-Coronado, M., et al.: Ecological role of volatile organic compounds emitted by pantoea agglomerans as interspecies and interkingdom signals. Microorganisms **9**(6), 1186 (2021)
2. Aziz, M., et al.: Augmenting sulfur metabolism and herbivore defense in arabidopsis by bacterial volatile signaling. Front. Plant Sci. **7**, 458 (2016)
3. Meldau, D.G., Meldau, S., Hoang, L.H., Underberg, S., Wünsche, H., Baldwin, I.T.: Dimethyl disulfide produced by the naturally associated bacterium bacillus sp b55 promotes nicotiana attenuata growth by enhancing sulfur nutrition. Plant Cell **25**(7), 2731–2747 (2013)
4. Farag El-Shafie, H.A., Faleiro, J.R.: Semiochemicals and their potential use in pest management. Biological control of pest and vector insects, pp. 10–5772 (2017)
5. Novoselov, K.S., et al.: Electric field effect in atomically thin carbon films. Science **306**(5696), 666–669 (2004)
6. Nayak, L., Rahaman, M., Giri, R.: Surface modification/functionalization of carbon materials by different techniques: an overview. Carbon-containing polymer composites, pp. 65–98 (2019)
7. Seitsonen, A.P., Marco Saitta, A., Wassmann, T., Lazzeri, M., Mauri, F.: Structure and stability of graphene nanoribbons in oxygen, carbon dioxide, water, and ammonia. Phys. Rev. B **82**(11), 115425 (2010)
8. Ge, H., Wang, G., Liao, Y.: Theoretical investigation on armchair graphene nanoribbons with oxygen-terminated edges. arXiv preprint arXiv:1411.1526 (2014)
9. Kostya S Novoselov, K.S., et al.: Two-dimensional gas of massless dirac fermions in graphene. Nature **438**(7065), 197–200 (2005)
10. Zhang, Y., Tan, Y.-W., Stormer, H.L., Kim, P.: Experimental observation of the quantum hall effect and berry's phase in graphene. Nature **438**(7065), 201–204 (2005)
11. Leenaerts, O., Partoens, B., Peeters, F.M.: Adsorption of h 2 o, n h 3, co, n o 2, and no on graphene: a first-principles study. Phys. Rev. B **77**(12), 125416 (2008)
12. Lin, X., Ni, J., Fang, C.: Adsorption capacity of h2o, nh3, co, and no2 on the pristine graphene. J. Appl. Phys. **113**(3), 034306 (2013)
13. Zhang, Z., et al.: Study on adsorption and desorption of ammonia on graphene. Nanoscale Res. Lett. **10**(1), 1–8 (2015)
14. Tarun, T., Singh, P., Kaur, H., Walia, G.K., Randhawa, D.K.K., Choudhary, B.C.: Defective gaas nanoribbon-based biosensor for lung cancer biomarkers: a DFT study. J. Mol. Modeling **27**(9), 270 (2021)
15. Walia, G.K., Randhawa, D.K.K., Singh, K., Singh, P., Kaur, H.: 25 detection of nitrous oxide gas using silicene nanoribbons. Intelligent Circuits and Systems, p. 147 (2021)
16. Kaur, H., Randhawa, D.K.K., Khosla, M., Sarin, R.K.: First principles study of sarin nerve gas adsorption on graphene nanoribbon with single molecule resolution. Mater. Today Proc. **28**, 1985–1989 (2020)
17. Aghaei, S.M., Monshi, M.M., Calizo, I.: Highly sensitive gas sensors based on silicene nanoribbons. arXiv preprint arXiv:1608.07508 (2016)
18. Quantum wise, copenhagen, denmark: Atomistix toolkit version2015.0 (2016). https://www.quantumwise.com. Accessed Jan-June 2019
19. Perdew, J.P., Burke, K., Wang, Y.: Erratum: generalized gradient approximation for the exchange-correlation hole of a many-electron system [phys. rev. b 54, 16 533 (1996)]. Phys. Rev. B **57**(23), 14999 (1998)

20. Abadir, G.B., Walus, K., Pulfrey, D.L.: Basis-set choice for DFT/NEGF simulations of carbon nanotubes. J. Comput. Electron. **8**, 1–9 (2009)
21. Deji, R., Choudhary, B.C., Sharma, R.K.: Novel hydrogen cyanide gas sensor: a simulation study of graphene nanoribbon doped with boron and phosphorus. Physica E: Low-dimensional Syst. Nanostructures **134**, 114844 (2021)
22. Monkhorst, H.J., Pack, J.D.: Special points for brillouin-zone integrations. Phys. Rev. B **13**(12), 5188 (1976)
23. Seenithurai, S., Pandyan, R.K., Kumar, S.V., Manickam, M.: Electronic properties of boron and nitrogen doped graphene. Nano Hybrids **5**, 65–83 (2013)

Author Index

© The Editor(s) (if applicable) and The Author(s), under exclusive license
to Springer Nature Switzerland AG 2023
M. K. Saini et al. (Eds.): ICA 2023, CCIS 1866, pp. 253–254, 2023.
https://doi.org/10.1007/978-3-031-43605-5

Printed in the United States
by Baker & Taylor Publisher Services